12th International Scientific Conference on Functional Materials and Nanotechnologies (FM&NT 2018)

IOP Conference Series: Materials Science and Engineering
Volume 503

Riga, Latvia
2 – 5 October 2018

ISBN: 978-1-5108-8487-8
ISSN: 1757-8981

Printed from e-media with permission by:

Curran Associates, Inc.
57 Morehouse Lane
Red Hook, NY 12571

Some format issues inherent in the e-media version may also appear in this print version.

Printed by Curran Associates, Inc. (2019)

For permission requests, please contact the Institute of Physics at the address below.

Institute of Physics
Dirac House, Temple Back
Bristol BS1 6BE UK

Phone: 44 1 17 929 7481
Fax: 44 1 17 920 0979

techtracking@iop.org

Additional copies of this publication are available from:

Curran Associates, Inc.
57 Morehouse Lane
Red Hook, NY 12571 USA
Phone: 845-758-0400
Fax: 845-758-2633
Email: curran@proceedings.com
Web: www.proceedings.com

Table of contents

Volume 503

Functional Materials and Nanotechnologies (FM&NT 2018)
2–5 October 2018, Riga, Latvia

Accepted papers received: 28 February 2019
Published online: 25 March 2019

Preface

Functional Materials and Nanotechnologies (FM&NT 2018)

Functional Materials and Nanotechnologies (FM&NT 2018)

Peer review statement

Papers

Negative thermal expansion of ScF_3: first principles vs empirical molecular dynamics

Dmitry Bocharov, Yuri Rafalskij, Matthias Krack, Mara Putnina and Alexei Kuzmin.....1

First principles evaluation on photocatalytic suitability of 2*H* structured and [0001] oriented WS_2 nanosheets and nanotubes

Y F Zhukovskii, S Piskunov and R A Evarestov.....5

Role of impurities in the formation of point defects in Ti-Al alloys

A V Bakulin, A A Fuks and S E Kulkova.....12

Influence of fluorine and oxygen atom adsorption on electronic structure of InSb(111) surface

A A Fuks, A V Bakulin and S E Kulkova.....16

Impedance spectroscopy of water splitting reactions on nanostructured metal-based catalysts

G A Niklasson, Z Qiu, I Bayrak Pehlivan and T Edvinsson.....20

TiO_2 thin films by ultrasonic spray pyrolysis

Z Chen, I Dündar, I Oja Acik and A Mere.....26

Deuterium concentration's in austenitic stainless steel by deuterium irradiation. Effects temperature irradiation

Oleksandr Morozov, Volodymyr Zhurba, Oleksandr Mats, Viktoria Progolaieva, Viktoriia Seliukova and Nikolay Kochnev.....33

Preparation of ZnO modified TiO_2 nanoporous coatings and their photocatalytic properties

R Drunka, J Grabis, A Letlena, Dz Jankovica, A Krumina and I Steins.....37

Electro-catalytic and photo-catalytic reformation of CO_2–reactions and efficiencies processes (Review)

A Knoks, P Lesnicenoks, J Kleperis, L Grinberga, J Hodakovska, J Klavins, G Cikvaidze and I Lukosevics.....42

Carbon nanotubes vs graphene nanoplatelets for 3D-printable composites

G. Gorokhov, D. Bychanok, D. Meisak, I. Shlyk, A. Liubimau, P. Angelova, C. Menseidov, E. Ivanov, R. Kotsilkova, M. Casa, P. Ciambelli and P. Kuzhir.....52

Wood-based nitrogen doped activated carbon for fuel cells

A Plavniece, G Dobele, A Volperts, A Zhurinsh and I Kruusenberg.....58

Electric field induced permittivity increase of carbon particle aggregates/silicone oil suspension

K Ozols, S Scegoleva and M Knite.....62

Synthesis and studying of reduced few-layered graphene coatings in gas sensor applications

Ingars Lukosevics, Peteris Lesnicenoks and Janis Kleperis.....66

Specific features of magnetic domain structure in epitaxial magneto-optical ferrite-garnet films

A I Ivanova, E M Semenova and M K Virchenko.....71

Determination of the electron effective mass for *n*-type germanium by the low-frequency impedance dispersion

O Malyshkina, E Barabanova, K Boitsova, V Klyuev and I Kaplunov.....75

Optical properties of zinc-iridium oxide thin films

V Skvortsova, M Zubkins, R Kalendarev, H Arslan and J Purans.....79

Electronic processes in doped ZnO nanopowders

Larisa Grigorjeva, Aleksejs Zolotarjovs, Ivita Bite, Jurgis Grube and Donats Millers.....83

Pressure-induced structural changes in α-MoO_3 probed by X-ray absorption spectroscopy

Inga Jonane, Andris Anspoks, Lucie Nataf, Francois Baudelet, Tetsuo Irifune and Alexei Kuzmin.....87

Nanoporous characterization of modified humidity-sensitive MgO-Al_2O_3 ceramics by positron annihilation lifetime spectroscopy method

H Klym, A Ingram, I Hadzaman, I Karbovnyk, I Vasylchyshyn and A I Popov.....91

Structural investigation of crystallized Ge-Ga-Se chalcogenide glasses

H Klym, O Shpotyuk, I Karbovnyk, L Calvez and A I Popov.....97

Luminescence characteristics of magnesium aluminate spinel crystals of different stoichiometry

G Prieditis, E Feldbach, I Kudryavtseva, A I Popov, E Shablonin and A Lushchik.....103

Float zone single crystals for testing rods, pulled under electron beam heating

Anatoly Kravtsov, Kirils Surovovs and Jānis Virbulis.....110

Influences of sputtering process factors on wear resistance of TiN coated on a machine component

Charnnarong Saikaew.....115

Features of nanostructure and functional properties formation in Ti-Ni shape memory alloys subjected to quasi-continuous equal channel angular pressing

I Yu Khmelevskaya, R D Karelin, S D Prokoshkin, M G Isaenkova, Yu A Perlovich, V A Fesenko, V S Komarov and M M Zaripova.....122

Analysis of the conditions of deformation's formation in lead selenide nanolayers

A M Pashaev, O I Davarashvili, M I Enukashvili, Z G Akhvlediani, L P Bychkova, R G Gulyaev, M A Dzagania and V P Zlomanov.....129

Optimization of the enzyme power source for a nano drug delivery system fuelled by glucose in blood plasma

S Naidoo, L Thage, Q Ying, S Vallie and G Vaivars.....136

Mass recovery of carbonated fabrics of glass fibres after isothermal heating

E Pentjuss, J Balodis, M Vdovicenko, J Gabrusenoks, G Bajars, L Petersone and J Kleperis.....142

Characterization of thermoelectric and thermogravimetric properties of conductive PEDOT:PSS films blended with SWCNTs and PVA

G Vugule, J Bitenieks, J Zicans, R Merijs-Meri, T Ivanova, D Erts and K Buks.....147

Model colloids to study surface – ligand interactions in nanosized Fe_3O_4

M M Maiorov, D Zablotsky, E Blums and A Krumina.....151

Different approaches in sulfonated poly (ether ether ketone) conductivity measurements

D Fedorenko and G Vaivars.....155

Nafion® and polyaniline composite modification with Li and Mg ions

Stanislavs Lozkins, Julija Hodakovska, Janis Kleperis and Remo Merijs-Meri.....160

IOP Conf. Series: Materials Science and Engineering **503** (2019) 011001 doi:10.1088/1757-899X/503/1/011001

Preface

The 12th International Scientific Conference 'Functional Materials and Nanotechnologies – FM&NT-2018' was held on October 2-5, in Riga, Radisson Blu Latvia Conference & Spa Hotel. The event was organized by the Institute of Solid State Physics, University of Latvia (ISSP UL). This conference is a continuation of meetings, first of which was organized in 2006 by ISSP UL in Riga. Since 2013 FM&NT conferences turned over a new page, becoming a common conference of all three Baltic countries. Now it is being organized periodically by the ISSP UL, University of Tartu, and Vilnius University. FM&NT-2018 is also a tribute to 40th anniversary of the ISSP UL, Latvia's Centenary programme Latvia 100 and beginning of Centenary year of the University of Latvia.

The purpose of the FM&NT is to bring together materials scientists, physicists, chemists, research staff, engineers, as well as experts in a wide range of the most demanding application areas, and students from universities, research institutes and related industrial companies. The topics of the conference included materials for energy and renewable energy technologies, biomedicine, multifunctional inorganic, organic and hybrid materials for photonics, micro- and nanoelectronics.

This year, more than 180 scientists from USA, Belarus, Denmark, South Africa, France, Georgia, Estonia, Italy, Israel, Kazakhstan, Russia, Lithuania, Norway, Poland, Portugal, Finland, Thailand, Taiwan, Ukraine, Hungary, Germany, Sweden, and, of course, Latvia presented their latest scientific achievements, reported on trends and activities in research and development of perspective materials using modern nanotechnology techniques.

6 plenary, 19 invited, 36 oral and 130 poster presentations reflected the results of theoretical and experimental research on functional materials and related processes, as well as possible application of innovative materials in science and economy.

Based on the conference contributions, 32 selected papers are included in this volume of *Journal of Physics Conference Series*: *Materials Science and Engineering*.

In additional meetings, cooperation in joint projects, the exchange of students and young researchers as well as governmental and political support for science were discussed. An overview of the ISSP UL co-ordinated project CAMART² was presented, which currently is the only 2nd phase project funded by the European Commission's Horizon 2020 Spreading Excellence and Widening Participation program in the Baltic Sea Region.

We would like to thank all the speakers, contributors, session chairs, referees and other involved staff for their efforts in making the FM&NT-2018 successful.

The Organizing and Program Committees are expressing their gratitude to Prof. George W.Taylor and Taylor&Francis, Inc. for the financial support and offer of a Special Integrated Ferroelectrics issue for the Proceedings of the conference.

Sincerely, organizers of the event
Andris Sternberg
Līga Grīnberga
Anatolijs Šarakovskis
Mārtiņš Rutkis

IOP Conf. Series: Materials Science and Engineering **503** (2019) 011001 doi:10.1088/1757-899X/503/1/011001

FM&NT-2018 International Organizing Committee

Chairman:

A. Sternberg (Latvia)

Members:

J. Banys (Lithuania), A. Bianconi (Italy), N. E.Christensen (Denmark), R. Evarestov (Russia), C.-G. Granqvist (Sweden), M. Kirm (Estonia), J. Kulda (France), E. Nõmmiste (Estonia), T. Plank (Estonia), M. Rutkis (Latvia)

Programme Committee

Chairman:

A. Šarakovskis

Members:

D. Erts, L. Grigorjeva, L. Grīnberga, M. Hammar, E. Kotomin, A. Kuzmin, A. Popov, J. Purāns, L. Skuja, K. Šmits, A. Sternberg, Q. Wang

Local Organizing Committee

Chairlady:

L. Grīnberga

Members:

A. Šarakovskis, J. Grube, E. Pavlovska, L. Klēbaha, A. Grūbe, I. Jonāne, I. Bite, A. Cintiņš, I. Lukoševičs, I. Takere, R. Siatkovskis

IOP Conf. Series: Materials Science and Engineering **503** (2019) 011002 doi:10.1088/1757-899X/503/1/011002

INSTITUTE OF SOLID STATE PHYSICS
UNIVERSITY OF LATVIA

CAMART²

FUNCTIONAL MATERIALS
& NANOTECHNOLOGIES

IOP Conf. Series: Materials Science and Engineering **503** (2019) 011002 doi:10.1088/1757-899X/503/1/011002

IOP Conf. Series: Materials Science and Engineering **503** (2019) 011003 doi:10.1088/1757-899X/503/1/011003

Peer review statement

All papers published in this volume of *IOP Conference Series: Materials Science and Engineering* have been peer reviewed through processes administered by the proceedings Editors. Reviews were conducted by expert referees to the professional and scientific standards expected of a proceedings journal published by IOP Publishing.

IOP Conf. Series: Materials Science and Engineering **503** (2019) 012001 doi:10.1088/1757-899X/503/1/012001

Negative thermal expansion of ScF_3: first principles vs empirical molecular dynamics

Dmitry Bocharov[1], Yuri Rafalskij[1], Matthias Krack[2], Mara Putnina[1] and Alexei Kuzmin[1]

[1]Institute of Solid State Physics, University of Latvia, Kengaraga street 8, LV-1063 Riga, Latvia

[2]Paul Scherrer Institut, Forschungsstrasse 111, 5232 Villigen PSI, Switzerland

E-mail: bocharov@latnet.lv

Abstract. The crystal lattice of cubic scandium fluorine (ScF_3) exhibits negative thermal expansion (NTE) over a wide range of temperatures from 10 K to 1100 K. Here the NTE effect in ScF_3 is studied using atomistic simulations based on empirical and *ab initio* molecular dynamics (AIMD) in the isobaric-isothermal (NpT) ensemble. The temperature dependence of the average lattice constant, the Sc–F–Sc bond angle distribution and the radial distribution functions were obtained. Crossover from the NTE to positive thermal expansion occurring at about 1100 K is reproduced by AIMD simulations in agreement with the known experiment data. At the same time, empirical MD model fails to reproduce the NTE behaviour and suggests an expansion of the ScF_3 lattice with increasing temperature. However, both MD models predict strong anisotropy of fluorine atom thermal vibration amplitude, being larger in the direction orthogonal to the Sc–F–Sc atom chain.

1. Introduction

Materials with a negative thermal expansion (NTE) coefficient were known since the beginning of the twentieth century but attracted the wider attention of researchers recently [1, 2, 3, 4]. The control of thermal expansion of materials is important in the technological areas, where the control of the volume of equipment components with high precision or the invariability of material volume with a change in temperature is required. This task can be solved, for example, by combining materials with positive and negative coefficients of thermal expansion (CTE). Possible areas of applications include electronics, fibre optics, astronomical and optical devices, dental prosthesis, etc [5].

Scandium fluoride (ScF_3) is remarkable among NTE materials due to it exhibits NTE properties in a very wide temperature range from 10 K to 1100 K and has isotropic and high CTE values [6]. Note that a structurally similar ReO_3 compound posses NTE effect with a very small CTE and below room temperature [7]. Unlike many other NTE materials as, for example, ZrW_2O_8, HfW_2O_8 [8] or $ZrV_{2-x}P_xO_7$ [9], ScF_3 has simple ABO_3 perovskite-like cubic lattice ($Pm\bar{3}m$ space group) with the A-site being empty that makes it a useful model object for studying NTE mechanism [10, 11, 12, 13, 14].

Possible mechanisms of the NTE effect were reviewed recently [2, 3, 4] and may have structural and electronic origin. Structural NTE is driven by transverse vibrational motion of the polyhedral units as happens in the case of ScF_3, whose structure is built-up by ScF_6

IOP Conf. Series: Materials Science and Engineering **503** (2019) 012001 doi:10.1088/1757-899X/503/1/012001

octahedra connected by vertices. Recently, Lasar et al. [15] showed that *ab initio* molecular dynamics (AIMD) can be used to model the NTE effect in ScF_3. They were able to reproduce temperature dependence of the reduced lattice constant, re-normalized at 200 K, using very small $2a_0 \times 2a_0 \times 2a_0$ (a_0 is the lattice constant of ScF_3) supercell.

In this study, we compare the results of empirical MD and AIMD simulations of ScF_3 performed for large supercells with the aim to check the limits of applicability of the two methods for a description of the NTE effect.

2. Computational details

AIMD simulations based on Kohn-Sham density functional theory (DFT) [16] with exchange-correlation potential PBEsol [17] and empirical MD simulations using the Born–Mayer potential [18, 19] were performed in the isobaric-isothermal (NpT) ensemble at seven different temperatures in the range of 300-1600 K using the CP2K code [20].

The equilibrium lattice constant of ScF_3 at T=0 K obtained from preliminary DFT calculations for optimized structure a_0=4.027 Å. The AIMD potential parameters were described previously in [21]. The Born–Mayer interatomic potential was used in empirical MD calculations

$$U(r) = A\exp(-r/B) \tag{1}$$

where A and B are potential parameters (A=942.4541 eV and B=0.2830 Å for the F–F atom pairs, whereas A=10173.51 eV and B=0.2003 Å for Sc–F atom pairs). No interaction between Sc atoms was considered. The potential parameters were optimized by fitting to the lattice constant and comparing the phonons given by this potential with phonon calculated in the Ref. [22]

The initial equilibration run was for both empirical MD and AIMD was 15 ps, followed by the production run of 50 ps. The Newton's equations of motion were integrated with the Verlet leapfrog algorithm, with a time step of 0.5 fs.

The temperature were controlled using the Nosé–Hoover thermostat. Sets of atomic configurations were generated during the MD simulations performed for $5a_0 \times 5a_0 \times 5a_0$ supercells with 500 atoms and used further to calculate the lattice constants a_0 (Fig. 1), the radial distribution functions (RDFs) Sc–F and Sc–Sc (Fig. 2) and the bond angle distribution functions (BADFs) Sc–F–Sc (Fig. 3).

3. Results and discussion

Temperature dependencies of the lattice constant a_0 obtained from empirical MD and AIMD simulations are compared with the experimental diffraction data from [6] in Fig. 1. The AIMD results reproduce the NTE effect in ScF_3 and agree well with the experimental data, overestimating only slightly (less than 0.01 Å) the absolute values of the experimental lattice constant. On the contrary, empirical MD predicts strong positive expansion of the lattice being always above both the experimental and AIMD data.

The difference in the behaviour of lattice constants for two simulation methods is also visible in the RDFs $G(r)$ in Fig. 2, where the RDF peaks obtained by empirical MD are slightly shifted to larger distances. However, both MD simulations predict anharmonic Sc–F bonding, leading to the visible asymmetry of the first peak in the RDFs at 2 Å. This result supports the importance of anharmonic effects in ScF_3 [10].

Finally, the temperature dependencies of the average bond angle Sc–F–Sc between two nearest ScF_6 octahedra are shown in Fig. 3. Note that there is a principal difference between the average bond angle Sc–F–Sc calculated from the instantaneous atomic coordinates, which deviates always from 180° due to transverse atom motion, and the Sc–F–Sc angle calculated from the average atomic positions, which is equal to 180° in cubic ScF_3. As one can see, an increase of temperature

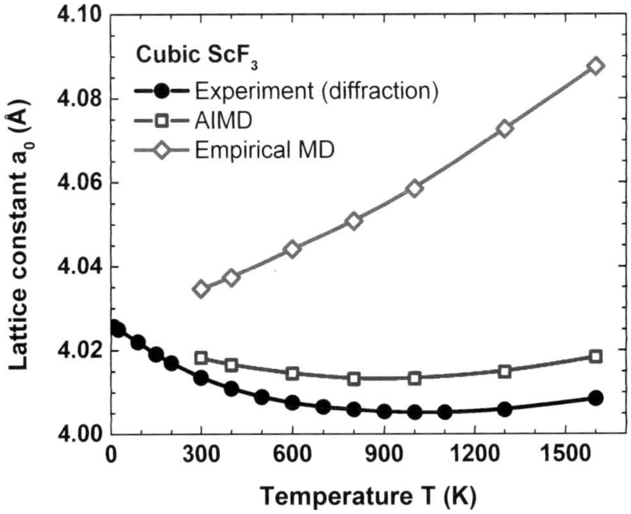

Figure 1. Temperature dependence of the lattice constant a_0 of cubic ScF_3 obtained from empirical MD and AIMD simulations in the NpT ensemble in comparison with the experimental data from [6].

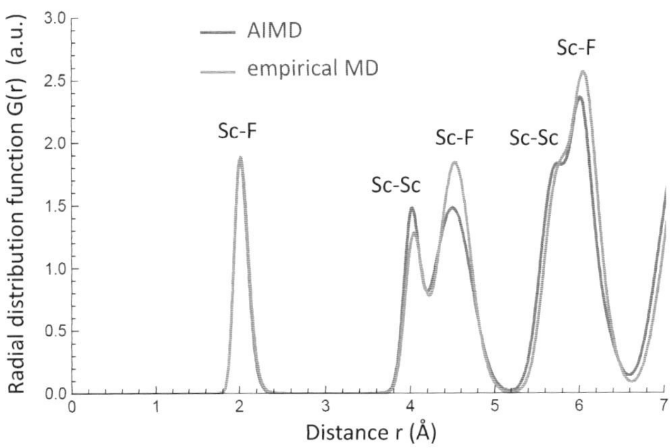

Figure 2. The total radial distribution functions $G(r)$ for Sc–F and Sc–Sc atom pairs at 600 K calculated using the results of the empirical MD and AIMD simulations in the NpT ensemble.

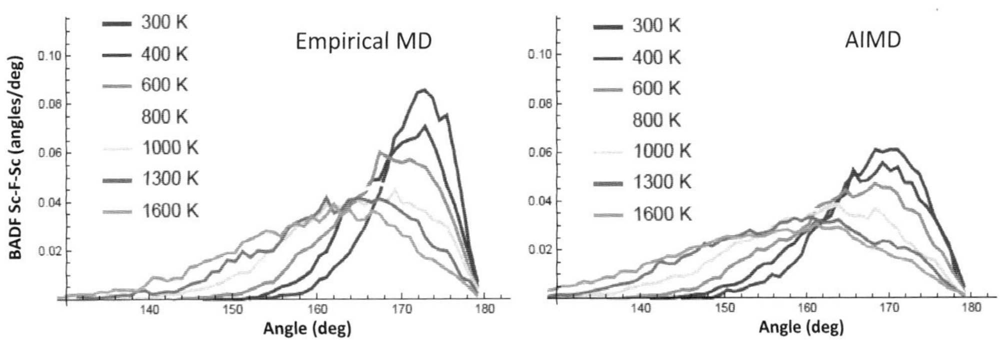

Figure 3. The inter-octahedral Sc–F–Sc bond angle distribution function calculated using the results of the empirical MD and AIMD simulations in the NpT ensemble.

leads to the broadening of BADF and its maximum shifts to smaller angles due to an increase of the amplitude of transverse vibrational motion of fluorine atoms. This effect is observed in both empirical MD and AIMD simulations. However, the Sc–F–Sc BADFs obtained by AIMD are always broader than those from empirical MD and extend to smaller angular values. This suggests slightly larger vibration amplitude of atoms in the AIMD simulations.

4. Conclusions

Ab initio and empirical molecular dynamics simulations of cubic ScF_3 were performed in the NpT ensemble and using large $5a_0\times5a_0\times5a_0$ supercells in the temperature range of 300-1600 K. Both simulations predict anharmonic nature of the Sc–F bonds and large amplitude of fluorine atoms transverse motion, which increases at higher temperatures. The negative thermal expansion of the ScF_3 lattice is well reproduced only by *ab initio* MD simulations, which predict the lattice constant values close to those obtained experimentally by diffraction in [6] and a crossover of thermal expansion from negative to positive around 1100 K upon increasing temperature. Empirical MD simulations fail to predict the NTE effect and shows only positive expansion.

Acknowledgments

The calculations were performed on the Paul Scherrer Institute cluster Merlin4, HPC resources of the Swiss National Supercomputing Centre in Lugano (project ID s626) as well as at the Latvian SuperCluster (LASC). Authors are greatly indebted to S. Ali, D. Gryaznov, R.A. Evarestov, M. Isupova, A. Kalinko, V. Kashcheyevs, V. Pankratov, S. Piskunov, A. I. Popov, J. Purans, F. Rocca, L. Shirmane, P. Žguns, and Yu. F. Zhukovskii for many stimulating discussions. Financial support provided by project No. 1.1.1.2/VIAA/l/16/147 (1.1.1.2/16/I/001) under the activity "Post-doctoral research aid" realized at the Institute of Solid State Physics, University of Latvia is greatly acknowledged.

References

[1] Lind C 2012 *Materials* **5** 1125–1154
[2] Dove M T and Fang H 2016 *Rep. Prog. Phys.* **79** 066503
[3] Mittal R, Gupta M and Chaplot S 2018 *Prog. Mater. Sci.* **92** 360–445
[4] Attfield J P 2018 *Front. Chem.* **6** 371
[5] Ventura G and Perfetti M 2014 *Thermal Properties of Solids at Room and Cryogenic Temperatures* (Springer)
[6] Greve B K, Martin K L, Lee P L, Chupas P J, Chapman K W and Wilkinson A P 2010 *J. Am. Chem. Soc.* **132** 15496–15498
[7] Dapiaggi M and Fitch A N 2009 *J. Appl. Crystallogr.* **42** 253–258
[8] Evans J S O, Mary T A, Vogt T, Subramanian M A and Sleight A W 1996 *Chem. Mater.* **8** 2809–2823
[9] Korthuis V, Khosrovani N, Sleight A W, Roberts N, Dupree R and Warren W W J 1995 *Chem. Mater.* **7** 412–417
[10] Piskunov S, Žguns P A, Bocharov D, Kuzmin A, Purans J, Kalinko A, Evarestov R A, Ali S E and Rocca F 2016 *Phys. Rev. B* **93** 214101
[11] van Roekeghem A, Carrete J and Mingo N 2016 *Phys. Rev. B* **94** 020303
[12] Bocharov D, Zguns P, Piskunov S, Kuzmin A and Purans J 2016 *Low Temp. Phys.* **42** 556–560
[13] Hu L, Chen J, Sanson A, Wu H, Guglieri Rodriguez C, Olivi L, Ren Y, Fan L, Deng J and Xing X 2016 *J. Am. Chem. Soc.* **138** 8320–8323
[14] Occhialini C A, Handunkanda S U, Curry E B and Hancock J N 2017 *Phys. Rev. B* **95** 094106
[15] Lazar P, Bučko T and Hafner J 2015 *Phys. Rev. B* **92** 224302
[16] Kohn W and Sham L J 1965 *Phys. Rev.* **140** A1133–A1138
[17] Perdew J P, Burke K and Ernzerhof M 1996 *Phys. Rev. Lett.* **77** 3865–3868
[18] Born M and Mayer J E 1932 *Eur. Phys. J. A* **75** 1–18
[19] Buckingham R A 1938 *Proc. R. Soc. London Ser. A* **168** 264–283
[20] CP2K developers group 2000–2018 (http://www.cp2k.org)
[21] Bocharov D, Krack M, Kalinko A, Purans J, Rocca F, Ali S E and Kuzmin A 2016 *J. Phys.: Conf. Ser.* **712** 012009
[22] Li C W, Tang X, Muñoz J A, Keith J B, Tracy S J, Abernathy D L and Fultz B 2011 *Phys. Rev. Lett.* **107** 195504

IOP Conf. Series: Materials Science and Engineering **503** (2019) 012002 doi:10.1088/1757-899X/503/1/012002

First principles evaluation on photocatalytic suitability of 2*H* structured and [0001] oriented WS_2 nanosheets and nanotubes

Y F Zhukovskii[1], S Piskunov[1] and R A Evarestov[2]

[1]Institute of Solid State Physics, University of Latvia, 8 Kengaraga Street, LV-1083 Riga, Latvia
[2]St. Petersburg State University, 7/9 Universitetskaya nab., 199034 St. Petersburg, Russia

e-mail: sergejs.piskunovs@cfi.lu.lv

Abstract. Pristine WS_2 multilayer nanosheets (NSs), which thickness h_{NS} varies from 1 to 12 monolayers (MLs), as well as single- and multi-walled nanotubes (SW and MW NTs) of different chirality, which diameter d_{NT} exceeds 1.9 nm, display photocatalytic suitability to split H_2O molecules. Obviously, such a phenomenon can occur since the band gap of these nanostructures corresponds to the energy range of visible light between the red and violet edges of spectrum (1.55 eV < $\Delta\varepsilon_{gap}$ < 2.65 eV). For all the studied WS_2 NSs and NTs, the levels of the top of the valence band and the bottom of the conduction band must be properly aligned relatively to H_2O oxidation and reduction potentials separated by 1.23 eV: $\varepsilon_{VB} < \varepsilon_{O_2/H_2O} < \varepsilon_{H^+/H_2} < \varepsilon_{CB}$. The values of $\Delta\varepsilon_{gap}$ decrease with growth of h_{NS} and increase with enlargement of d_{NT}. 1 ML nanosheet can be considered as a limit of infinite SW NT thickness growth ($d_{NT}\rightarrow\infty$), which band gap increases up to ~2.65 eV. First principles calculations have been performed using the hybrid DFT-HF method (HSE06 Hamiltonian) adapted for 2*H* WS_2 bulk. The highest solar energy conversion efficiency (15-18%) expected at $\Delta\varepsilon_{gap}$ = 2.0-2.2 eV (yellow-green range) has been found for 2 ML thick (stoichiometric) WS_2 (0001) NS as well as WS_2 NTs with diameters 2.7-3.2 nm (irrespectively on morphology and chirality indices *n* of nanotubes). Moreover, unlike discrete variation of h_{NS} magnitudes, tuning of d_{NT} values provides much higher energy resolution.

1. Introduction

Photocatalytic water splitting under influence of solar light in vicinity of semiconducting electrode immersed in aqueous medium is a clean and renewable source for the production of hydrogen fuel [1], which plays essential role in contemporary energetics. A number of semiconductors of various chemical nature, dimensionality and morphology were studied previously as potential candidates for photocatalytic electrodes, including transition metal dichalcogenides (TMDC) attracting enhanced attention during last 10-15 years due to the progress in their fabrication and application [2]. The optimum band gap of semiconducting electrode for dissociation of H_2O molecules caused by the influence of visible light lies in the range of 2.0–2.2 eV [3]. Absorption of photons results in migration of excited electrons over the bottom of conduction band (CB), which simultaneously creates holes under the top of valence band. Since centers of dissociation located on the catalyst surface are important for the water splitting in order to prevent recombination between electrons and holes [4], reproducible fabrication of nanostructures is one of ways how to ensure the direct exit of recombination centers to the surface, thus noticeably enhancing photocatalytic efficiency. First of all 2D nanothin films, 1D nanotubes and nanoribbons as well as 0D fullerene-like nanocages and nanoclusters can be applied for photocatalytic splitting of water. In the current study, we analyze periodic nanosystems only.

Morphology of TMDC bulk materials can be presented in the form of 1*T*, 2*H* and 3*R* phases corresponding to trigonal, hexagonal as well as rhombohedral crystalline lattices described by D_{3d}, D_{3h} and C_{3v}^5 point groups, respectively [5]. These types of crystals establish the class of two-periodic layered structures characterized by strong chemical (mainly covalent) bonds within the single monolayers (MLs) and weak coupling between the neighboring MLs. Of particular interest is their high catalytic activity towards the hydrogen evolution reaction [6], which could support efficient and low-cost production of hydrogen without reliance on expensive Pt-based catalysts. The most widespread two-periodic transition metal dichalcogenides applied for photocatalytic splitting of water and other molecules were found to be MoS_2 and WS_2 disulfides [7], which prefer to be stacked in *2H c* bulk phase and can be "restacked" from separate graphene-like monolayers when using the corresponding experimental methods [8].

When applying the layered TMDC materials engineering, various experimental techniques (*e.g.*, atomic force microscopy and optical contrast) provide precise control of perfectness and thickness of nanothin films deposited on supporting substrates [9]. Atomically thin WS_2 nanosheets were prepared applying sulfurization of the ultrathin WO_3 films [10] by adjusting the number of plasma enhanced atomic layer deposition during the process cycles. The high uniformity of mono-, bi-, and tetra-layered WS_2 films synthesized upon α-Al_2O_3 and SiO_2/Si substrates was achieved when using the low-pressure chemical vapor deposition (CVD) method [11, 12] which is predominantly applied until now. Mechanically exfoliated WS_2 flakes consisting of a few monolayers covered the transparent quartz or SiO_2/Si substrates with incorporated iodine (providing abundant active edges for photocatalytic reactions) were carefully studied using CVD method too [13]. Nanothin films consisting of a few WS_2 monolayers were also experimentally observed to grow upon [0001]-oriented ZnO nanowires [14], formation of which was confirmed by using the methods of transmission electron microscopy, Raman scattering and optical spectroscopy. Photoluminescence of WS_2 nanosheets, as reported recently [15], weakens with increasing number of monolayers due to a transition from the direct band gap in a monolayer to the indirect gap in multilayers.

To determine the band structure of WS_2 single crystal and monolayer, both augmented spherical wave calculations based on density functional theory (DFT) and angle-resolved photoelectron spectroscopy using synchrotron radiation source were applied [16]. Comprehensive DFT plane wave study of the electronic structure and photocatalytic properties for a number of TMDC single MX_2 monolayers was performed recently [17] using many-body G_0W_0 Green function approach and either GGA PBE [18] or hybrid HSE06 functionals [19]. In order to estimate the proper alignment of the top of valence band ε_{VB} and the bottom of conduction band ε_{CB} with the oxidation and reduction potentials (ε_{O_2/H_2O} and ε_{H^+/H_2}, respectively) as a criterion of photocatalytic suitability, the quasiparticle band gap was constructed relatively to its center following the method proposed by Toroker *et al* [20].

For the first time, R. Tenne *et al.* synthesized WS_2 nanotubes (NTs) in 1992 [21] following the later intensive experimental studies on them. For example, a successful attempt to synthesize Nb-doped WS_2 NTs was undertaken [22], which inhibits the formation of defective nanotubes unlike routine synthesis of pristine nanotubes. Controlled Re-doping of WS_2 nanotubes and fullerene-shaped structures resulted in a drop of the electrical resistivity, leading to a more efficient electron transport through the corresponding nanotubes [23]. Combination of metallic Co nanoparticles and semiconducting WS_2 NTs results in essential enhancement of the photoactivity of hybrid nanostructures [24]. Therefore, efficient light absorption by nanotubes can occur and subsequent charge separation between the semiconducting WS_2 nanotube and the metallic nanocluster takes place under visible light illumination.

However, theoretical characterization of WS_2 NTs is still in its infancy [25, 26]. The latter took into account approaches based on the line symmetry group formalism, using either valence force-fields combined with semi-empirical density functional tight binding theory (DFTB) [25] or an *ab initio* hybrid DFT-HF formalism using the HSE06 functional in which the DFT Hamiltonian is extended by inclusion of Hartree-Fock (HF) exchange and correlation [26]. Recent *ab initio* calculations have focused on comparison between the electronic structures in WS_2 bulk as well as its layered and nanotubular forms [27]. In those publications and additionally in Ref. [28] both vibrational and electronic properties of WS_2 NTs were calculated and analyzed, however, not their possible photocatalytic applications.

To fill the gap existing in comprehensive investigations of WS_2 monoperiodic structures, especially their photocatalytic suitability for water splitting, we have evaluated recently various WS_2 nanosheets [9] and nanotubes [29] performing for this aim large-scale *ab initio* calculations on their electronic structure. In this study, we compare the corresponding results to estimate which efficiency is higher.

2. Theoretical background

2.1. *Hybrid DFT&HF-LCAO method*

Large scale *ab initio* calculations on [0001]-oriented WS_2 nanosheets and nanotubes with varied thickness h_{NS} and diameter d_{NT}, respectively, have been performed using the hybrid density functional theory (DFT)/Hartree-Fock (HF) calculations using the *CRYSTAL14* computer code [30]. It utilizes localized Gaussian type functions (GTFs) in the form of atom-centered basis sets (BSs) for expansion of periodic crystalline orbitals as linear combinations of atomic orbitals (LCAO). We have used the HSE06 hybrid functional [19], which proved to be the most adequate for reproducing the experimental value of band gap for tungsten disulfide bulk (1.58 eV *vs.* 1.4 eV obtained in our fully relaxed calculations [9] and experimental measurements [31], respectively). Balanced summation in both direct and reciprocal WS_2 nanosheets and nanotubes requires choice of the reciprocal space integration by sampling the Brillouin zone (BZ) with 18×18×1 and 18×1×1 Monkhorst–Pack meshes [32], respectively, which result in total of 37 and 10 *k*-points evenly distributed over the corresponding BZs. For the fixed crystalline geometry, calculations converge when the total energy differs by less than 10^{-9} a. u. in two successive cycles of the self-consistent field (SCF) procedure [30]. Full geometry optimization has been performed for all considered in this study WS_2 bulk, slab and tubular models.

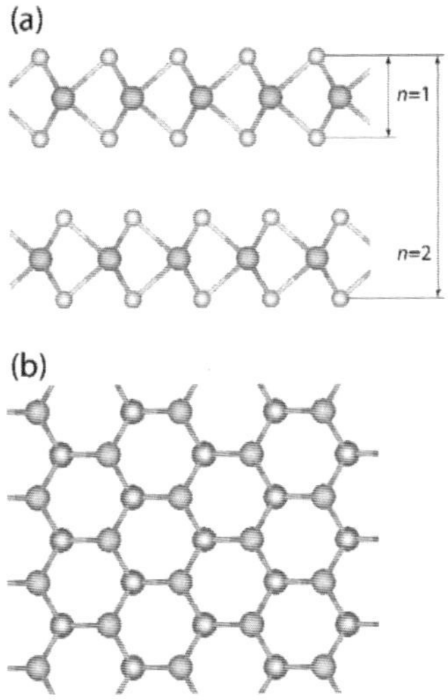

Figure 1. Models of single- and double-layered WS_2 (0001) nanosheets described by hexagonal symmetry of *2H c* phase (aside (a) and atop (b) views). W and S atoms are marked with large green (dark) balls and small yellow (light) balls, respectively.

2.2. *Models of WS_2 nanosheets and nanotubes*

The primitive unit cell of hexagonal *2H c* phase mentioned above as the most stable for WS_2 bulk consists of two atomic inversely arranged layers containing three atomic planes each [14]. Inner planes of each WS_2 layer contain only tungsten atoms while in its turn outer planes contain sulfur atoms. Due to weak interaction forces acting between monolayers, as considered above, the inter-layer distance, which corresponds to a half of the lattice constant *c* for bulk tungsten disulphide) is essentially larger (6.16 Å) as compared to the constant *a* (3.15 Å) [14]. In order to check the photocatalytic suitability of graphene-type WS_2 (0001) nanosheets with varied thicknesses h_{NS}, we have constructed stacks of 1-12 MLs (corresponding to 0.31-5.91 nm thickness) and, additionally, models containing up to 20, 30, and 40 MLs. Models of tungsten disulfide nanothin films consisting of single- and double-layered are shown in Figure 1. Nanosheets containing odd (1, 3, 5, . . ., $2n$+1 MLs) and even (2, 4, 6, . . ., $2n$ MLs) numbers of WS_2 monolayers, where $n = 0$, 1, 2, . . ., possess two different layer symmetry groups $P\bar{6}m2$ and $P\bar{3}m1$, respectively.

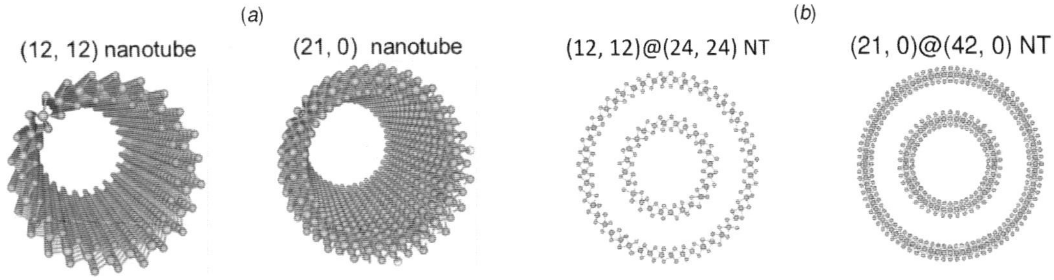

Figure 2. Models of SW (*a*, axonometric images) and DW (*b*, cross-sections perpendicular to NT axes) [0001]-oriented WS_2 nanotubes possessing armchair (n, n) and zigzag-type (n, 0) chiralities. W and S atoms are marked with large dark balls and small light balls, respectively.

Construction of [0001] oriented WS_2 NTs described by folding the corresponding nanolayer results in formation of the cylindrical nanotubes by rolling up the graphene-structured nanosheets using the formalism of line groups [25, 26]. For this purpose the nomenclature of single-walled (m, n) carbon nanotubes can be expressed *via* the so-called chirality indices m and n as described earlier [33]. To clarify dependence of photocatalytic suitability on NT morphology, we have performed *ab initio* calculations on WS_2 NTs with both achiral single-walled armchair (n, n) and zigzag-type $(n, 0)$ structure and diameters $d_{NT} > 1.0$ nm (Figure 2a) as well as double-walled $(m, m)@(n, n)$ and $(m, 0)@(n, 0)$ nanotubes, internal diameters of which exceed 1.0 nm (Figure 2b).

The achiral armchair (n, n) and zigzag $(n, 0)$ single-walled tungsten disulphide nanotubes with hexagonal morphology belong to 4 and 8 line-group families with $L(2n)_n/m$ and $L(2n)_n\,mc$ symmetry, accordingly [26]. Both types of WS_2 NTs with achiral morphology possess the helical axes of order $2n$.

3. Results of calculations, their analysis and comparison

3.1. Suitability of periodic pristine WS_2 nanostructures for photocatalytic applications

Several special conditions are required for a material to be photocatalytically active for water splitting. First, the band gap of the catalyst must correspond to the visible light range between 1.5-2.7 eV. For the solar intensity distribution, the maximum efficiency is achievable for a band gap between 2.0 and 2.2 eV. The energy conservation condition requires that both the oxidation and the reduction potentials of the H_2O molecule in solution defined thermodynamically (ε_{O_2/H_2O} = -5.67 eV and ε_{H^+/H_2} = - 4.44 eV) [1] must be positioned inside the band gap of the photocatalytic electrode. Thus, the following inequalities have to be hold:

$$\varepsilon_{VB} < \varepsilon_{O_2/H_2O} < \varepsilon_{H^+/H_2} < \varepsilon_{CB}. \qquad (1)$$

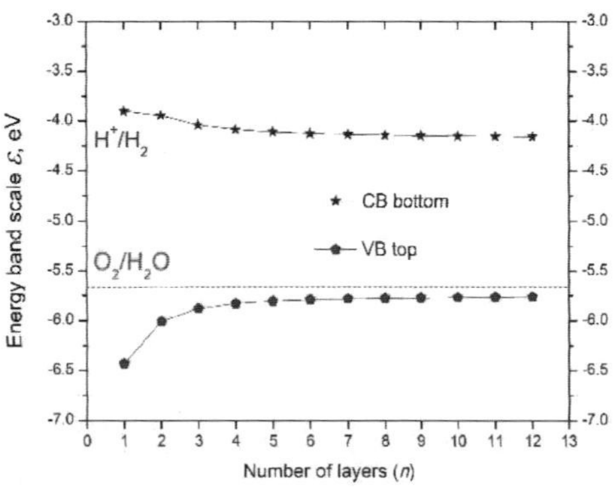

Figure 3. Dependence of WS_2 (0001) band gap parameters on nanolayer thickness with alignment of ε_{CB} and ε_{VB} edges relative to the redox potentials.

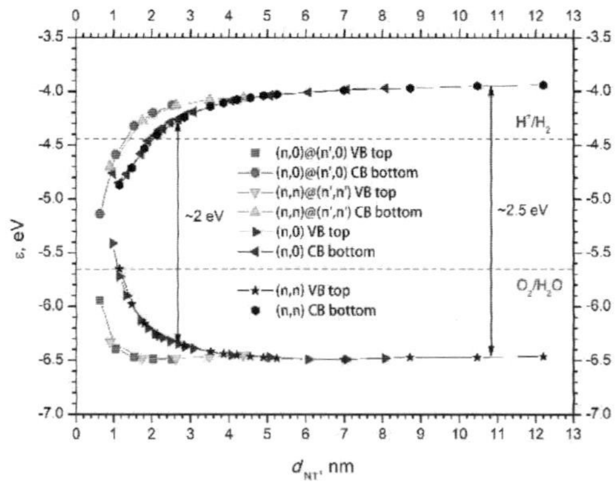

Figure 4. Dependence of band gap parameters for SW and DW WS_2 NTs on the values of d_{NW} (inner for DW) with alignment of ε_{CB} and ε_{VB} edges relative to ε_{O_2/H_2O} and ε_{H^+/H_2} levels.

Figures 3 and 4 present results of our calculations on the band edge parameters of twelve WS_2 nanosheets and approximately forty different nanotubes, respectively, including the positions of ε_{O_2/H_2O} and ε_{H^+/H_2} at the energy scale. One can see that levels of ε_{VB} for all NSs and NTs are always below the ε_{O_2/H_2O} level; thus, the left part of Eq. (1) is satisfied for all nanotubes. In contrary, if for all nanosheets the right part of Eq. (1) is always satisfied, then for thin NTs ($d_{NT} < 1.9$ nm), the level of ε_{CB} is located below the ε_{O_2/H_2O}, which prevents the catalytic decomposition of H_2O molecules. However, further increase of d_{NT} leads to a rapid shift of the ε_{CB} level upward, thus, Eq. (1) is valid for all nanotubes, diameter of which exceeds 1.9 nm, irrespective of chirality. We note that the increase of the band gap of nanosheets with reduced thickness can be attributed to the quantum confinement effect, while in nanotubes the strain effect is predominant and therefore the bandgap shrinks with reducing nanotube's diameter.

3.2 Comparison of band structure for double- and monoperiodic pristine WS_2 nanostructures

We have performed recently detailed analysis of DOSs for WS_2 nanosheets [9] and nanotubes [29]. In this study, we analyze and compare the band structure of tungsten disulphide nanosheets and two achiral nanotubes, respectively. In Figure 5, we compare the band structures of WS_2 (0001) monolayer and bilayer using Γ(0, 0, 0) – M(1/2, 0, 0) – K(1/3, 1/3, 1/3) – Γ(0, 0, 0) paths for both in the Brillouin Zone. In the case of 1 ML, the optical direct band gap is located at K-point. For bilayers and more thick nanosheets, we observe indirect gaps only. When the number of monolayers increases, the fundamental band gap decreases and tends to bulk value.

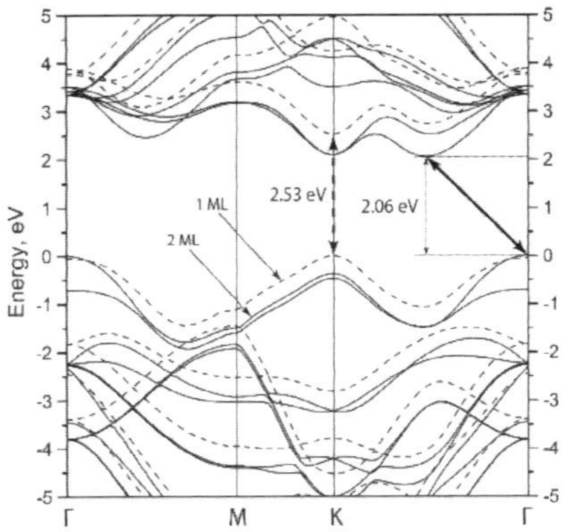

Figure 5. Band structure for WS_2 NSs consisting of 1 ML (dashed lines) and 2 ML (solid lines).

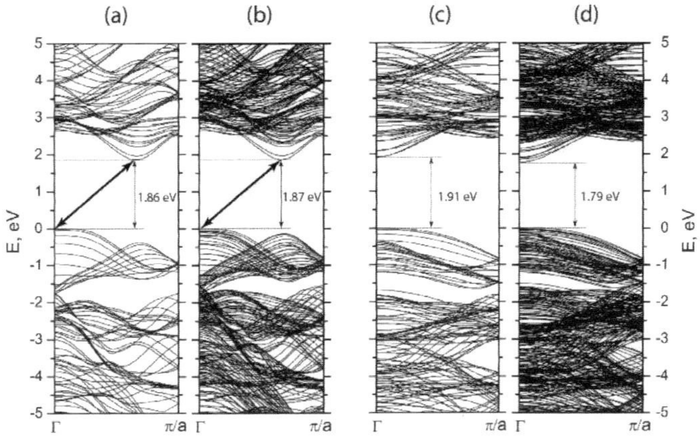

Figure 6. Band structure for WS_2 NTs possessing armchair and zigzag-type chiralities: (a) SW (12,12) WS_2 NT, (b) DW (12,12)@(24,24) WS_2 NT, (c) SW (21,0) WS_2 NT, and (d) DW (21,0) @(42,0) WS_2 NT.

In the case of both achiral SW and DW WS_2 NTs (Figure 6), the indirect band gaps characterize armchair (*ac*) chirality (a, b), while for zigzag-type (*zz*) nanotubes, our calculations yield direct Γ– Γ band gap, irrespectively to the number of NT walls (c, d). These results contradict to conclusion drawn for band structures of single- and double-layered WS_2 (0001) nanosheet (Figure 5) characterized by direct and indirect transitions from the bottom of the CB to the top of the VB, respectively.

4. Conclusions

We have simulated the electronic structure of pristine WS_2 multilayer nanosheets with varied thickness as well as single- and double-walled WS_2 nanotubes of different chirality with varied diameter. 2D-peiodic nanostructures are found suitable for photocatalytic applications since their band gaps correspond to the whole range of visible light between the red and violet edges (1.5 eV < $\Delta\varepsilon_{gap}$ < 2.7 eV) while the top of valence band and the bottom of conduction band are properly aligned relative to the oxidation and reduction potentials, *i.e.*: $\varepsilon_{VB} < \varepsilon_{O_2/H_2O} < \varepsilon_{H^+/H_2} < \varepsilon_{CB}$.

The only limitation for photocatalytic suitability of WS_2 NTs is determined by d_{NT} > 1.9 nm (for DW NTs, this condition concerns the internal wall). On the other hand, tuning of d_{NT} provides much higher energy resolution than in the case of nanosheets. Moreover, energies of $\Delta\varepsilon_{gap}$ edges depend only on diameter d_{NT} (increasing monotonously with increasing diameter) and do not depend on chirality indices *n*. Principal difference between both types of WS_2 nanostructures is that the value of $\Delta\varepsilon_{gap}$ for nanotubes grows with increasing d_{NT}, while that of nanosheets decreases with enlargement of h_{NS}. Furthermore, the bandgap of DW NT is smaller than the corresponding$\Delta\varepsilon_{gap}$ values for each of its constituents.

The highest conversion efficiency of single-walled nanotubes is achieved between the yellow-green ranges of visible spectrum ($2.0 < \Delta\varepsilon_{gap} < 2.2$ eV), which corresponds to diameter variation 2.7 nm $< d_{NT}$ < 3.2 nm, irrespectively on chirality. For WS_2 (0001) nanosheets, the highest photocatalytic efficiency is achieved for 2 ML thickness due to discreteness of h_{NS} values, thus giving a preference to nanotubes.

Acknowledgements
This study was supported by the EC ERA.Net RUS Plus Project No. 237 WATERSPLIT. R.E. acknowledges the financial support provided by the Russian Foundation for Basic Research (grant N 17-03-00130a) and High Performance Computer Center of St. Petersburg University for the assistance. The authors are indebted to D. Bocharov, O. Lisovski and E. Spohr for stimulating discussions.

References
[1] Chen X, Shen S, Guo L and Mao S S 2010 *Chem. Rev.* **110** 6503
[2] Dong R and Kuljanishvili I 2017 *J. Vacuum Sci. Technol. B* **35** 030803
[3] Navarro R M, Alvarez-Galvan M C, Villoria de la Mano J A, Al-Zahrani S M and Fierro J L G 2010 *Energy Environ. Sci.* **3** 1865
[4] Kudo A and Miseki Y 2009 *Chem. Soc. Rev.* **38** 253
[5] Toh R J, Sofer Z, Luxa J, Sedmidubsky D and Pumera M 2017 *Chem. Commun.* **53** 3054
[6] Benck J D, Hellsternn T R, Kibsgaard J, Chakthranont P and Jaramillo T F 2014 *ACS Catal.* **4** 3957
[7] Chen T E, Chang Y H, Hsu C L, Wei K H, Chiang C Y and Li L J 2013 *Int. J Hydrogen Energy* **38** 12302
[8] Petkov V, Billinge S J L, Heising J and Kanatzidis M G 2000 *J. Am. Chem. Soc.* **122** 11571
[9] Bocharov D, Piskunov S, Zhukovskii Y F and Evarestov R A 2018 *phys. status solidi rrl* **12** 1800253
[10] Zeng W. Feng L-P, Su J, Pan H-xi and Liu Z-T 2018 *J. Alloys&Compd.* **745** 834
[11] Zhang Y, Zhang Y, Ji Q, Ju J, Yuan H, Shi J, Gao T, Ma D, Liu M, Chen Y, Song X, Hwang H Y, Cui Y and Liu Z 2013 *ACS Nano* **7** 8963
[12] Park J, Lee W, Choi T, Hwang S-H, Myoung J M, Jung J-H, Kim S-H and Kim H 2015 *Nanoscale* **7** 1308
[13] Zhao W, Ghorannevis Z, Chu L, Toh M, Kloc C, Tan P-H and Eda G 2013 *ACS Nano* **7** 791
[14] Polyakov B, Kuzmin A, Smits K, Zideluns J, Butanovs E, Butikova J, Vlassov S, Piskunov S and Zhukovskii Y F 2016 *J. Phys. Chem. C* **120** 2145
[15] Gutiérrez H R, Peres-López N, Elias A L, Berkdemir A, Wang B, Lv R, López-Urias F, Crespi V, Terrones H and Terrones M 2013 *Nano Lett.* **13** 3447
[16] Klein A, Tiefenbacher S, Eyert V, Pettenkofer C and Jaegermann W 2001 *Phys. Rev. B* **64** 205416
[17] Zhuang H L and Hennig R G 2013 *J. Phys. Chem. C* **117** 20440
[18] Perdew J P, Burke K and Ernzerhof M 1996 *Phys. Rev. Lett.* **77**, 3865
[19] Heyd J, Scuseria G E, and Ernzerhof M 2003 *J. Chem. Phys.* **118** 8207
[20] Toroker M C, Kanan D K, Alidoust N, Isseroff L Y, Liao P and Carter E A 2011 *Phys. Chem. Chem. Phys.* **13** 16644
[21] Tenne R, Margulis L, Genut M and Hodes G 1992 *Nature* **360** 444
[22] Zhu Y Q, Hsu W K, Firth S, Terrones, M, Grobert N, Clark R J H, Kroto H W, Walton D R M 2001 *Chem. Phys. Lett.* **342** 15
[23] Yadgarov L, Rosentsveig R, Leitus G, Albu-Yaron A, Moshkovich A, Perfilyev V, Vasic R, Frenkel A I, Enyashin A N, Seifert G, Rapoport and Tenne R 2012 *Angew. Chemie Intl. Ed. Engl.* **51** 1148
[24] Tsverin Y, Livneh T, Rosentsveig R, Zak A, Pinkas I and Tenne R. 2013 *Nanomater. Energy* **2** 25
[25] Damnjanović M, Vuković T and Milošević I 2017 *Isr. J. Chem.* **57** 450
[26] Evarestov R A, Bandura A V, Porsev V V and Kovalenko A V 2017 *J. Comput. Chem.* **38** 2581
[27] Zibouche N, Kuc A and Heine T 2012 *Eur. Phys. J. B* **85** 1

[28] O'Neal K R, Cherian J G, Zak A, Tenne R, Liu Z and Musfeldt J L 2016 *Nano Lett.* **16** 993
[29] Piskunov S, Lisovski O, Zhukovskii Y F, Dyachkov P N, Evarestov R A, Kenmoe S and Spohr E 2018 *ACS Omega* **4** 1434
[30] Dovesi R, Saunders V R, Roetti C, Orlando R, Zicovich-Wilson C M, Pascale F, Civalleri B, Doll K, Harrison N M, Bush I J, D'Arco Ph, Llunell M, Causá M and Noël Y 2014 *CRYSTAL14 User's Manual*, University of Torino
[31] Braga D, Gutiérrez Lezama I, Berger H and Morpurgo A F 2012 *Nano Lett.* **12** 5218
[32] Monkhorst H J and Pack J D 1976 *Phys. Rev. B* **13** 5188
[33] Saito R, Dresselhaus M S and Dresselhaus G 1998 *Physical Properties of Carbon Nanotubes*.(Imperial College Press, World Scientific Publishing Co. Pte. Ltd., Singapore)

IOP Conf. Series: Materials Science and Engineering **503** (2019) 012003 doi:10.1088/1757-899X/503/1/012003

Role of impurities in the formation of point defects in Ti-Al alloys

A V Bakulin[1,2], A A Fuks[1] and S E Kulkova[1,2]

[1] National Research Tomsk State University, pr. Lenina 36, Tomsk 634050, Russia
[2] Institute of Strength Physics and Materials Science of the Russian Academy of Sciences, Siberian Branch, pr. Akademichesky 2/4, Tomsk 634055, Russia

E-mail: bakulin@ispms.tsc.ru

Abstract. The effect of impurities on the point defect formation energies in Ti-Al alloys was studied by the projector augmented-wave method. The grand canonical formalism was applied to calculate the formation energies of several kinds of defects. It was shown that the formation energy of Al-vacancy decreases by ~1.3 eV with increasing its content in the series of alloys Ti_3Al–TiAl–$TiAl_3$, whereas the formation energy of Ti-vacancy changes insignificantly. In general, the formation of Ti-vacancies is more preferred in Ti-Al alloys rather than the formation of Al-vacancies. We demonstrate that Nb, Mo, Tc, Ru, Rh, Pd lead to increase of the formation energy of Al-vacancy if they occupy the Al sites, whereas only Mo and Tc increase the formation energy of Ti-vacancy. All 4*d* elements, if they substitute for Ti, reduce the formation energies of Al-vacancy. The influence of impurities on elastic constants is discussed.

1. Introduction

Intermetallic Ti-Al alloys are very attractive as high-temperature structural materials for aerospace, automotive, marine and other applications. These alloys have unique mechanical properties such as a high specific strength, advanced creep characteristics and relatively good oxidation/corrosion resistance. The development of new alloys based on these compounds, whose properties are between nickel-based superalloys and advanced ceramics, is a challenging problem of modern materials science. One of the methods that allows one to improve the functional properties of alloys is their doping. Atomic diffusion mechanisms and mechanical properties of alloys strongly depend on impurities. It is well-known that all self-diffusion mechanisms require the existence of vacancies or complexes of point defects. Although the formation energy of different defects and their complexes in binary alloys has been intensively studied, this topic remains in focus of numerous theoretical studies [1]. The formation energies of some defects in TiAl and Ti_3Al alloys were calculated earlier in [2] using embedded-atom potentials. It was shown that in both alloys antistructural defects strongly dominate over vacancies. However, the information about the influence of impurities on the defect formation energy in Ti-Al alloys is insufficient. In this work we study the effect of 4*d* impurities and Ga, Ge, In, Sn, etc. on the point defect formation energies in the series of alloys Ti_3Al–TiAl–$TiAl_3$.

2. Method of calculation

The calculations of atomic and electronic structure of binary alloys were performed by the projector augmented-wave method (PAW) method [3] implemented in the VASP code [4]. The generalized gradient approximation (GGA-PBE) [5] was used for the exchange-correlation functional. The full optimization including volume and shape of crystal structures was carried out until the forces at atoms were smaller than 0.01 eV/Å. A *k*-point mesh of 7×7×7 obtained according to the Monkhorst–Pack scheme was used for integration over the Brillouin zone in (2×2×2) supercell of γ-TiAl and in (2×2×1) supercell of $TiAl_3$. In case of Ti_3Al, the (2×2×2) supercell was used alongside with the 3×3×4 Γ-centered *k*-point mesh. The formation energy of defects was calculated using the following equation:

$$E_f = E_{def} - E_{id} + n_A\mu_A + n_B\mu_B \tag{1}$$

where E_{def} and E_{id} are the total energies of the supercell with defect and the ideal one (the difference between these energies (E^{raw}) is called "raw" energy), μ_A and μ_B – the chemical potentials of binary alloy $A_\alpha B_\beta$ components, n_A and n_B – the number of atoms of A and B added to ($n < 0$) and/or removed ($n > 0$) from the perfect crystal to form the defect. In order to calculate E_f, the approach suggested in [6] for an alloy of equiatomic composition was used.

3. Results and discussion

The main problem of the calculations of defect formation energies is connected with the determination of the chemical potentials of the alloy components. In most works the chemical potentials of bulk constituent elements are used. Another approach for the estimation of μ_A and μ_B was suggested in [6], in which the grand canonical formalism was used. It was shown that the chemical potential of component i = A, B in the linear approximation can be expressed as follows

$$\mu_i = \mu_i^0 - T\tilde{\mu}_i(x), \tag{2}$$

where x is a deviation of the alloy composition from the stoichiometric ($A_{\alpha+x}B_{\beta-x}$), μ_i^0 is the chemical potential at zero temperature, $\tilde{\mu}_i(x)$ is the coefficient of the linear temperature dependence of the chemical potential, T – temperature. Since antisites are the constitutional defects in Ti-Al alloys, their concentrations are much higher than those of the vacancies. The following expressions for μ_i^0 and $\tilde{\mu}_i$ were obtained in case of $Ti_{\alpha+x}Al_{\beta-x}$ alloy by modification of the equations (A3-A6, A11-A14) from [6]:

$$\text{for Ti-rich } (x>0) \begin{cases} \mu_{Ti}^0 = \varepsilon_0 + \beta\, E^{raw}(Ti_{Al}),\ \tilde{\mu}_{Ti} = \beta\, k_B \ln(1/2x), \\ \mu_{Al}^0 = \varepsilon_0 - \alpha\, E^{raw}(Ti_{Al}),\ \tilde{\mu}_{Al} = -\alpha\, k_B \ln\left(4(0.5-x)^2/2x\right) \end{cases} \tag{3}$$

and

$$\text{for Al-rich } (x<0) \begin{cases} \mu_{Ti}^0 = \varepsilon_0 - \beta\, E^{raw}(Al_{Ti}),\ \tilde{\mu}_{Ti} = -\beta\, k_B \ln\left(-4(0.5+x)^2/2x\right) \\ \mu_{Al}^0 = \varepsilon_0 + \alpha\, E^{raw}(Al_{Ti}),\ \tilde{\mu}_{Al} = \alpha\, k_B \ln(-1/2x) \end{cases}, \tag{4}$$

where ε_0 is the total energy per atom for the ideal supercell, $E^{raw}(Ti_{Al})$ and $E^{raw}(Al_{Ti})$ are the raw energies of Ti and Al antisites, respectively. In case of the stoichiometric composition, the chemical potential $\tilde{\mu}_i$ equals 0 and μ_i^0 is calculated as an average of two values for $x > 0$ and $x < 0$ [2].

The obtained values of the formation energies of vacancies and antisites for Ti_3Al, TiAl and $TiAl_3$ alloys are shown in Figure 1. It is seen that vacancies have higher formation energies than antisites in all considered compounds. The difference in E_f is the largest in Ti_3Al alloy but it decreases with increase of Al content. In general, the defect formation energies are more sensibly affected by the alloy composition in $TiAl_3$ than in both TiAl and Ti_3Al alloys. It is seen from Figure 1 that Ti vacancies form easier than Al ones in the alloys of the stoichiometric composition. The difference in E_f for Al and Ti vacancies in stoichiometric alloy TiAl reaches 0.39 eV that is in agreement with the value of 0.41 eV [7] obtained by *ab-initio* method, but it is slightly larger than that of 0.25 eV obtained by EAM [2]. At the same time, the calculated values of E_f in case of Ti-vacancy in $Ti_{48.5}Al_{51.5}$ and $Ti_{65.6}Al_{34.4}$ alloys are 1.63 eV and 2.10 eV that are larger than corresponding experimental values

1.41±0.06 eV [8] and 1.55±0.2 eV [9] obtained by the positron lifetime spectroscopy for the alloys of the same compositions and at the same temperature. It is known that *ab-initio* methods within DFT overestimate the defect formation energy in alloys. However, the trend in increase of the formation energy of Ti-vacancy with increase of Ti content in alloy is in line with the experiment.

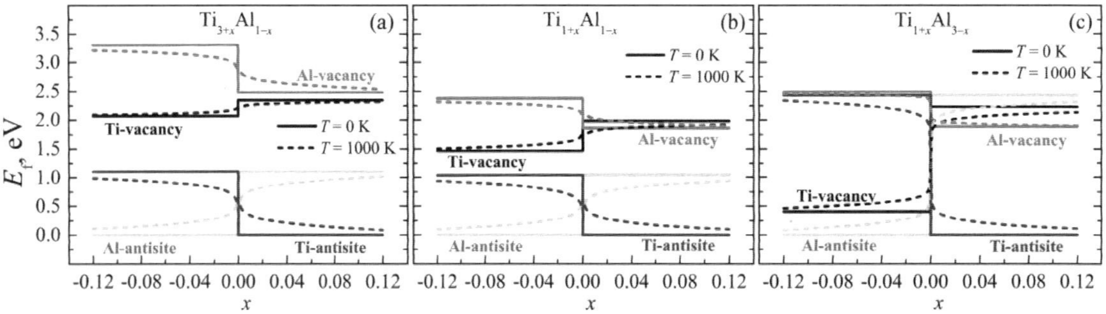

Figure 1. Formation energies of point defects (vacancies and antisites) in Ti_3Al (a), TiAl (b) and $TiAl_3$ (c) as functions of composition near the stoichiometry.

The influence of impurities on E_f in TiAl alloy is summarized in Figure 2. Note that the impurities were located in the first neighbors of defects. It is interesting that impurities such as Nb and Mo increase the formation energy of all point defects if they occupy Al-sublattice but this effect is less pronounced in case of Al antistructural defect. The impurities of groups IIIA and IVA (Si, Ga, Ge, In, Sn) on both Al and Ti sites provide an Al removal from their local environment. All these impurities lead also to reduce of E_f for Al-vacancies and Ti-antisites but they increase of Al-antisite formation energy. Impurities of Ru, Rh, Pd and Ag substituting for Al atom affect the similar way in respect to titanium. It should be noted that a dramatic downfall of the vacancy formation energy at the addition of Y and Zr instead of Al is connected with a displacement of the impurity atoms from their positions towards vacancy. As a result, the final structure looks like complex of the interstitial impurity atom and two vacancies. All considered impurities, if they occupy the Ti sites, provide the decrease of the formation energy of Al-vacancy. Finally, it is necessary to emphasized that the site substitution behavior is found to be in agreement with earlier experimental and theoretical results [7,10,11].

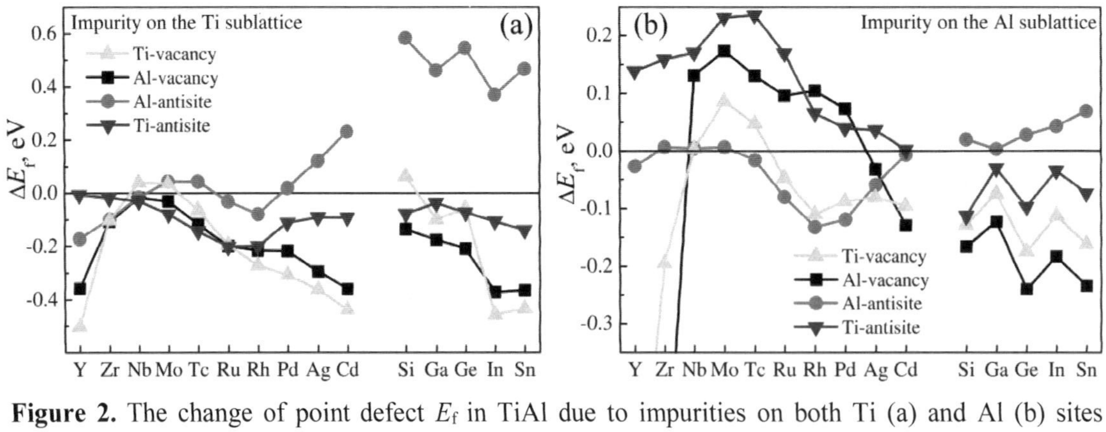

Figure 2. The change of point defect E_f in TiAl due to impurities on both Ti (a) and Al (b) sites relative to undoped alloy.

Several words should be said about influence of impurities on the elastic moduli of γ-TiAl alloy. It is seen from Table 1 that calculated C_{ij} are in good agreement with earlier available theoretical and experimental results. Note that the local density approximation for the exchange-correlation functional and pseudopotential approach was used in [7]. Modulus C_{44}, which reflects the resistance to shear in the (001) plane in the [010] or [100] direction, agrees well with the experimental data, whereas other moduli demonstrate the deviations as compared to the other theoretical results. It is almost twice larger than the corresponding one in Ti_3Al (55.5 GPa). Modulus C_{66} related to the resistance to shear in the (010) plane in the [100] direction, also is smaller than that for Ti_3Al (51.1 GPa). So, the Ti-rich alloy is elastically softer than γ-TiAl against shear modes. Impurities such as Nb, Mo and also Zr, Tc, Ru, Rh on Ti sites lead to increase of bulk modulus that is in line with conclusion from [10], whereas Young's and shear moduli are decreased in doped alloy excepting the case of Mo. At the same time, Ga and Ge affect the *B* modulus slightly and their effects on *E* and *G* are opposite.

Table 1. Calculated elastic (C_{ij}), bulk (*B*), Young's (*E*) and shear (*G*) moduli (in GPa) for undoped and doped γ-TiAl alloy

	C_{11}	C_{12}	C_{13}	C_{33}	C_{44}	C_{66}	*B*	*E*	*G*
γ-TiAl	168.0	89.2	90.4	171.8	113.3	84.1	116.4	175.4	70.2
[7]	177	99	93	181	111	–	125	–	–
exp. [12]	187	75	75	182	109	81.2	111	–	–
Nb on Ti site	167.3	96.7	89.5	178.8	114.0	79.7	118.3	174.3	69.5
Mo on Ti site	171.0	100.6	89.2	183.0	113.4	82.3	120.3	177.0	70.5
Ga on Al site	168.0	91.8	89.9	172.5	112.6	73.2	116.9	171.2	68.2
Ge on Al site	171.4	92.3	87.3	183.1	112.1	76.0	117.7	177.0	70.9

4. Conclusion

In summary, the influence of impurities on the formation energies of point defects and elastic moduli in Ti-Al alloys was performed within DFT method. We demonstrate that the formation energy of Al-vacancy depends stronghly on Al content in the Ti-Al alloy, whereas the formation energy of Ti-vacancy changes insignificantly on alloy composition. In spite of all 4*d* impurities lead to decrease of the formation energies of Al-vacancy, if they substitute for Ti, only impurities of the beginning of 4*d* period prefer to be on Ti-sites. Besides, Nb and Mo increase the formation energy of Ti-vacancies. We established that impurities such as Tc, Ru, Rh, Pd, which prefer to occupy the Al sites, lead to increase of the formation energy of Al-vacancies. The calculations of elastic moduli allow to confirm that TiAl is stiffer than Ti_3Al against shear modes. Impurities such as Nb and Mo, if they substitute for Ti, and also simple metals on the Al-sites lead to increase of bulk modulus.

Acknowledgment

The work was partly supported by the Russian Foundation for Basic Research under Grant 18-03-00064_a, ISPMS SB RAS (project III.23.2.8) and Tomsk State University Competitiveness Improvement Program. Numerical calculations were performed on the supercomputers at TSU and MSU.

References

[1] Rudolph P 2008 Thermodynamics, origin, and control of defects *Crystal Growth Technology* ed H Scheel, P Capper (Weinheim: Wiley VCH) pp. 73–102
[2] Mishin Y and Herzig Chr 2000 *Acta Mater.* **48** 589
[3] Blöchl P E 1994 *Phys. Rev.* B **50** 17953
[4] Kresse G and Joubert J 1999 *Phys. Rev.* B **59** 1758
[5] Perdew J P, Burke K and Ernzerhof M 1996 *Phys. Rev. Lett.* **77** 3865
[6] Mayer J, Elsasser C and Fahnle M 1995 *Phys. Stat. Sol. B* **191** 283
[7] Woodward C, Kajihara S and Yang L H 1998 *Phys. Rev. B* **57** 13459
[8] Brossmann U, Wurschum R, Badura K and Schaefer H-E 1994 *Phys. Rev. B* **49** 6457
[9] Würschum R, Kümmerle E A, Badura-Gergen K, Seeger A, Herzig Ch and Schaefer H-E 1996 *J. Appl. Phys.* **80** 724
[10] Hao Y L, Xu D S, Cui Y Y, Yang R and Li D 1999 *Acta Mater.* **47** 1129
[11] Music D and Schneider J M 2006 *Phys. Rev. B* **74** 174110
[12] Tanaka K, Ichitsubu T, Iniu H, Yamaguchi M and Koiwa M 1996 *Philos. Mag. Lett.* **73** 71

IOP Conf. Series: Materials Science and Engineering **503** (2019) 012004 doi:10.1088/1757-899X/503/1/012004

Influence of fluorine and oxygen atom adsorption on electronic structure of InSb(111) surface

A A Fuks[1], A V Bakulin[1,2] and S E Kulkova[1,2]

[1] National Research Tomsk State University, pr. Lenina 36, Tomsk 634050, Russia
[2] Institute of Strength Physics and Materials Science of the Russian Academy of Sciences, Siberian Branch, pr. Akademichesky 2/4, Tomsk 634055, Russia

E-mail: artemfuchs45@gmail.com

Abstract. The projector augmented-wave method was applied for investigation of oxygen and fluorine atom adsorption on the InSb(111) surface depending on its termination. It was shown that oxygen adsorption leads to appearance of additional surface states in the fundamental gap in case of the In-terminated surface while the density of surface states decreases for the Sb-terminated surface. Fluorine co-adsorption results in partial or complete removal of the surface state and an unpinning of the Fermi level. The increase of fluorine concentration leads to considerable changes of the near-surface-layer structure due to the penetration of both electronegative adsorbates (O and F) into the substrate.

1. Introduction

Semiconductors based on elements of III and V groups are promising materials for various device applications due to their superior electron mobility, as compared to silicon. The progress in manufacturing of devices based on $A^{III}B^{V}$ semiconductors is connected with the development of a special technique, which provides the low density of states (DOS) at insulator–semiconductor interfaces caused by the formation of native oxide on a semiconductor surface [1]. Controlled growth of fluorine-containing anodic oxide layers (FAOL) in different media is one of the ways to modify a semiconductor surface. For example, this method provides low DOS ($<5\times10^{10}$ eV^{-1} cm^{-2}) at the interfaces Au/FAOL/InAs(111)A and In_2O_3/SiO_2/FAOL/InAs(111)A [2]. The letter A in $A^{III}B^{V}$(111)A means that the surface terminates by atomic layer of A atoms. In case of GaAs the formation of FAOL/GaAs interface leads to decrease of DOS by 50 times in comparison with the interface formed without fluorine [3]. The complex composition of the FAOL/$A^{III}B^{V}$ interface transition region considerably complicates finding out the microscopic nature of the interface states.

The investigations of the FAOL/InAs(111)A interface morphology carried out earlier by the high-resolution electron microscopy, which showed the formation of a well-ordered crystalline interface and a transition region of four to six atomic layers thickness [2]. Thus, the structure of the semiconductor at the interface is rather ideal than reconstructed one. In this connection *ab-initio* methods, which allow to understand the origin of the interface states and the chemical bonding of an adsorbate on a semiconductor surface, are actively used. Earlier the initial stage of formation of interface between FAOL and InAs(111)A, GaAs(111)A substrates was investigated in [4,5]. It was shown that the fluorination of indium and arsenic oxides, forming the native oxide layers on the (111) surface leads to radical changes of the atomic structures of the subsurface region. The oxygen-induced surface states are complete or partial removed from the band gap if fluorine forms bonds with the indium atoms involved in an interaction with oxygen. At the same time, the question of generality of this surface passivation mechanism for $A^{III}B^{V}$ semiconductors remains under debate. In the present work, in order to understand the FAOL/InSb interface formation mechanism, we study the influence of oxygen and fluorine atom adsorption and their co-adsorption, and also fluorine concentration (up to three atoms) on the electron properties of the InSb(111) unreconstructed surface.

2. Method of calculation

The atomic and electronic structures of the InSb(111) unreconstructed surface were calculated using the projector augmented-wave [6,7] method (PAW) implemented in the VASP code. The generalized gradient approximation within PBE form [8] for the exchange-correlation functional, as well as the hybrid functional HSE06 [9] was employed. The surface was simulated by repeated slabs separated by a vacuum region of ~10 Å in thickness between them. The slab consists of eight atomic layers of a semiconductor and a layer of pseudo-hydrogen atoms with fractional number of valence electrons. One surface was saturated with pseudo-hydrogen atoms to passivate dangling bonds; therefore, one film side was a bulk-like. On the opposite film side fluorine and oxygen atom adsorption was considered. Two indium and antimony layers from the side of the hydrogen-saturated film were fixed at bulk positions, whereas atoms of other layers can relax during the structure optimization. The relaxation of atomic positions was carried out until the forces at atoms ~0.01 eV/Å. A plane-wave cutoff of 500 eV was used throughout. Integration over the Brillouin zone was carried out using a grid of *k*-points (7×7×1) obtained according to the Monkhorst-Pack scheme. Since the (111) surface is a polar one, dipole corrections were taken into account. Electronic states were considered to be surface states if virtually all of their charge was contained within the top few surface layers, whereas bulk states have most of their charge confined to deeper layers. The theoretical lattice parameters of bulk InSb (6.652 Å), obtained within the PAW-PBE approximation, are slightly larger than the experimental ones, 6.479 Å [10]. The binding energies of oxygen and fluorine atoms with the InSb substrate were calculated using the equation (2) from [4].

3. Results and discussion

The calculation of the atomic structure of InSb(111)-(1×1) unreconstructed surface demonstrates the inward relaxation of upper atomic layer irrespective of surface termination. The relaxation of the first interlayer distance Δ_{12} is equal to -5.5% in case of the In-terminated InSb(111)A surface, whereas it is -2.1% for InSb(111)B surface. It should be noted that smaller values in magnitude (-3.8% and -5.4%) were obtained for InAs(111)A and GaAs(111)A surfaces [4]. This indicates a shorter distance between cation and arsenic layers on the surface, which is due to the decrease of surface cation coordination. Our calculations show that the relaxation of interlayer distances has oscillation character and decreases into the film depth. Electronic structure calculations demonstrate that in case of clean surfaces there are surface states in the fundamental band gap, which are pinning the Fermi level (Figure 1). These

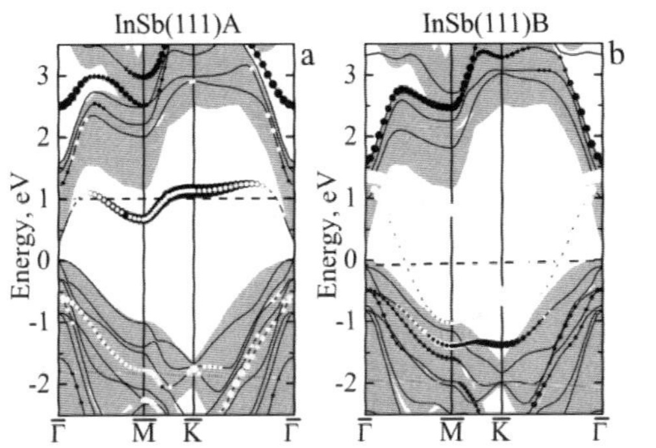

Figure 1. Electronic band structure along high-symmetry directions of Brillouin zone for InSb(111) with the In (a) and Sb (b) terminations.

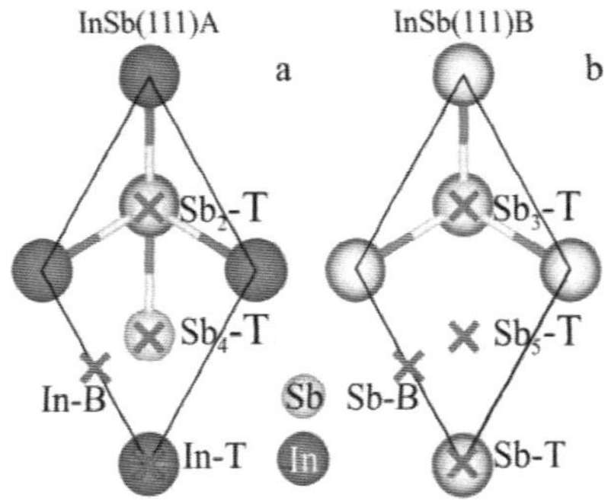

Figure 2. Atomic structure of the InSb(111) surface with the In (a) and Sb (b) terminations and the adsorbate positions (crosses).

states spread over the whole two-dimensional Brillouin zone. On the In-termination the surface states are conditioned by surface In and subsurface Sb atoms while in case of the Sb-termination they are connected with Sb surface atoms. In both cases these surface states are formed by the dangling *p*-orbitals. It should be noted that the black and yellow circles in Figure 1 are indium and antimony surface states, and their size is proportional to the localization degree of electronic states on the surface atoms. We remind that the electronic structure of the surface was calculated within HSE06 method and the band gap was found to be 0.254 eV, which agrees with the experimental value of 0.235 eV [11]. However, a band gap value of zero was obtained within the PAW-PBE calculation.

The adsorption of oxygen and fluorine atoms was considered at high-symmetry sites shown in Figure 2. The binding energies on both surfaces are presented in Table 1. It is seen that fluorine has the largest binding energy in the top T-site above the indium surface atom, whereas oxygen prefers to be adsorbed in the bridge B-site on both surface terminations. The preference of the latter site is connected with the fact that oxygen has bonds with two In surface atoms (the bond length is ~2.40 Å) and additionally with Sb atom of the subsurface layer (the bond length is ~2.06 Å). Note that the similar trend was observed earlier on InAs(111)A and GaAs(111)A surfaces [4]. It is interesting that the oxygen binding energies with the substrate change slightly when it is adsorbed at the sites above antimony atoms of the second (third) and fourth (fifth) layers on the In(Sb)-terminated InSb(111) surface. In all cases the Bader charge of oxygen atom is larger than that of fluorine. It is seen that the interaction of both adsorbates with the surface has ionic character irrespective of its termination.

Table 1. Binding energies (in eV) and the Bader charges (in *e*, in brackets) of F and O adatoms on the both InSb(111)-(1×1) surface terminations.

Surface	Adatom	In(Sb)-B	In(Sb)-T	$Sb_2(Sb_3)$-T	$Sb_4(Sb_5)$-T
InSb(111)A	F	4.54 (0.76)	5.14 (0.59)	4.22 (0.75)	4.42 (0.79)
	O	5.93 (1.04)	4.35 (0.65)	4.98 (1.02)	4.84 (1.00)
InSb(111)B	O	6.05 (1.01)	5.30 (0.70)	5.78 (0.97)	5.81 (0.97)

Oxygen adsorption on both terminations of InSb(111) surface results in significant changes of surface states (Figure 3). In case of the In-terminated surface the oxygen adsorption leads to the appearance of additional states in the band gap and, as it can be seen from Figures 3a and 3b, the structure of these surface states depends strongly on an adsorbate site on the surface. The opposite trend is observed on the Sb-terminated surface (Figures 3c and 3d). In this case the surface states are unoccupied and they are significantly shifted towards the conduction band due to adsorption of oxygen. As the result the Fermi level crosses the states which are mainly conditioned by bulk-like atoms. Such changes of the surface electronic structure at the first step of oxidation are in line with experimental observations: density of states at the interface between InSb and native oxide is smaller in case of the Sb-terminated surface in comparison with the In-terminated one [12].

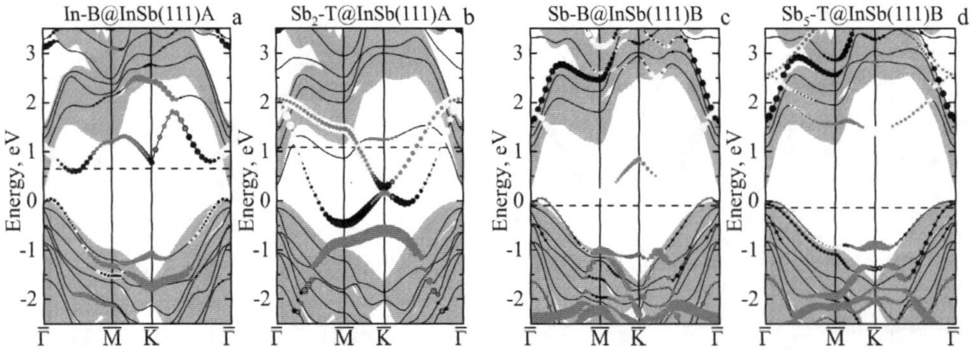

Figure 3. Electronic band structure of the InSb(111) surfaces with oxygen adsorbed in In-B (a) and Sb_2-T (b) sites on the In-termination as well as with oxygen adsorbed in Sb-B (c) and Sb_5-T (d) sites on the Sb-terminated surface.

Finally, let's discuss the changes in the surface electronic structure of InSb(111)A due to oxygen and fluorine co-adsorption. It is seen in Figure 4a that fluorine co-adsorption leads to decrease of the density of surface states in the band gap. The oxygen-induced surface states are practically complete removed by the co-adsorption of three F atoms (Figure 4b): only the states near the band gap edges remain. In this case oxygen and two fluorine atoms lie in the subsurface region (Figure 4c) that can be considered as initial stage of interface formation between InSb substrate and fluorine-containing anodic oxide. The increase of fluorine concentration at more than three atoms per (1×1) surface cell leads to the appearance of additional states near top of the valence band due to their partial depletion.

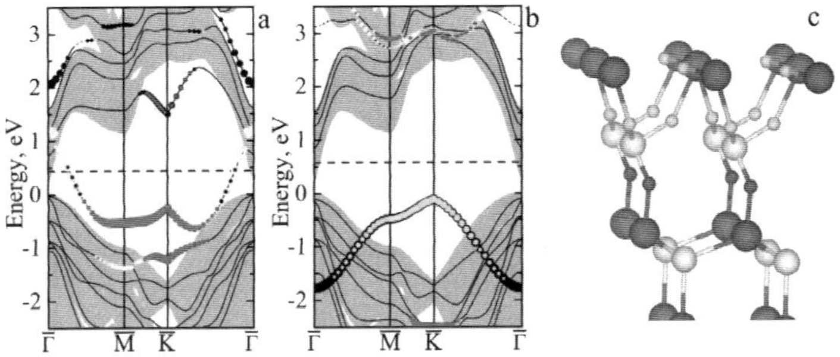

Figure 4. Electronic band structure of the InSb(111)A surfaces with oxygen adsorbed in the In-B site and with one (a) and three (b) fluorine atoms as well as corresponding atomic structure (c).

4. Conclusion

The influence of fluorine and oxygen adsorption and their co-adsorption on the surface states in the band gap was studied on both In- and Sb-terminated InSb(111) surface. It was shown that in former case oxygen results in appearance of additional surface states, while density of surface states in the gap decreases sharply in case of the Sb-termination. The oxygen-induced surface states on InSb(111)A are complete or partial removed from the band gap by fluorine co-adsorption if it forms bonds with the indium atoms involved in an interaction with oxygen. The fluorine co-adsorption and an increase of its concentration up to three atoms result in significant structural changes in the subsurface region due to the penetration of both oxygen and fluorine atoms into substrate that leads to breaking of semiconductor bonds in the near-surface region.

Acknowledgment

The work was partly supported by ISPMS SB RAS (project III.23.2.8) and Tomsk State University Competitiveness Improvement Program. The calculations were performed on the TSU supercomputer.

References

[1] Hinkle C L, Vogel E M, Ye P D and Wallace R M 2011 *Curr. Opin. Solid State Mater. Sci.* **15** 188
[2] Aksenov M S, Kokhanovskii A Yu, Polovodov P A et al. 2015 *Appl. Phys. Lett.* **107** 173501
[3] Ahrenkiel R K, Kazmenski L L, Ireland P J, Jamjoum O, Russell P E, Dunlavy D, Wagner R S, Pattillo S and Jervis T 1982 *J. Vac. Sci. Technol.* **21** 434
[4] Bakulin A V, Kulkova S E, Aksenov M S and Valisheva N A 2016 *J. Phys. Chem.* C **120** 17491
[5] Valisheva N A, Bakulin A V, Aksenov M S, Khandarkhaeva S E and Kulkova S E 2017 *J. Phys. Chem.* C **121** 20744
[6] Blöchl P E 1994 *Phys. Rev.* B **50** 17953
[7] Kresse G and Joubert J 1999 *Phys. Rev.* B **59** 1758
[8] Perdew J P, Burke K and Ernzerhof M 1996 *Phys. Rev. Lett.* **77** 3865
[9] Heyd J and Scuseria G E 2004 *J. Chem. Phys.* **121** 1187
[10] Sze S M 1981 *Physics of Semiconductor Device* (New York: Wiley Interscience Publication)
[11] Vurgaftman I, Meyer J R and Ram-Mohan L R 2001 *J. Appl. Phys.* **89** 5815
[12] Meiners L G 1985 Electrical Properties of Insulator-Semiconductor Interfaces on III-V Compounds *Physics and Chemistry of III-V Compound Semiconductor Interfaces* ed C W Wilmsen (New York: Plenum Press) pp 213–281

IOP Conf. Series: Materials Science and Engineering **503** (2019) 012005 doi:10.1088/1757-899X/503/1/012005

Impedance spectroscopy of water splitting reactions on nanostructured metal-based catalysts

G A Niklasson, Z Qiu, I Bayrak Pehlivan and T Edvinsson

Department of Engineering Sciences, Uppsala University, P.O. Box 534, SE-75121 Uppsala, Sweden

Corresponding author: gunnar.niklasson@angstrom.uu.se

Abstract. Hydrogen production by water splitting using nanomaterials as electrocatalysts is a promising route enabling replacement of fossil fuels by renewable energy sources. In particular, the development of inexpensive non-noble metal-based catalysts is necessary in order to replace currently used expensive Pt-based catalysts. We report a detailed impedance spectroscopy study of Ni-Mo and Ni-Fe based electrocatalytic materials deposited onto porous and compact substrates with different conductivities. The results were interpreted by a critical comparison with equivalent circuit models. The reaction resistance displays a strong dependence on potential and a lower substrate dependence. The impedance behaviour can also provide information on the dominating reaction mechanism. An optimized Ni-Fe based catalyst showed very promising properties for applications in water electrolysis.

1. Introduction

Hydrogen is a promising alternative to fossil fuels and can also be used as energy storage. In particular, it is useful for storing electricity from intermittent renewable sources, which later can be used to produce heat, or alternatively electricity on-demand via fuel cells [1]. Hydrogen can be produced by electrocatalysis of a hydrogen-containing compound such as water. In order to lower the overpotential (i.e., the excess above the thermodynamic requirement of 1.229 V), highly efficient catalysts for the hydrogen evolution reaction (HER) and oxygen evolution reaction (OER) are needed [2]. Currently used highly effective Pt- and Ir-based electrocatalysts need to be replaced by less expensive and more earth-abundant alternatives and here transition metal alloys, oxides/hydroxides or other transition metal compounds are of prime interest [3,4].

The HER is quite well understood in both acidic and alkaline solutions. In the latter case, it is summarized by the following basic reaction steps [5-7]:

$$M + H_2O + e^- \leftrightarrow M - H + OH^- \quad \text{(Volmer step)} \quad (1)$$
$$M - H + H_2O + e^- \leftrightarrow H_2 + OH^- + M \quad \text{(Heyrovsky step)} \quad (2)$$
$$M - H + M - H \leftrightarrow H_2 + 2M \quad \text{(Tafel step)} \quad (3)$$

Here M represents the usually metallic active sites on the surface of the catalyst. It is realized that hydrogen can be produced either by reaction pathway (1) and (2) or by (1) and (3). The OER is considerably more complicated, both in number of possible reaction paths and number of possible steps involved [8,9].

Information on rate-limiting reaction steps and reaction kinetics can be obtained from analysis of so-called Tafel plots depicting the close to exponential dependence of the charge transfer current on overpotential [10], or from frequency-dependent measurements using impedance spectroscopy [6,11]. In the present paper, we focus on the impedance spectroscopy technique and present impedance spectra pertaining to the HER for two transition metal-based electrocatalysts prepared by different methods. The data are interpreted in terms of equivalent circuits, both empirical ones and those that can be derived from the theory of the HER. We especially discuss the substrate dependence of the impedance parameters and discuss the promising properties of a Ni-Fe based catalyst.

2. Experimental methods

Ni-Mo films were prepared by reactive DC magnetron co-sputtering using a Balzers UTT 400 unit. The targets were metallic Ni (99.99% purity, Plasmaterials) and Mo (99.99% purity, Plasmaterials) discs. The target to substrate distance was 13 cm. Depositions were done in Ar gas at 30 mTorr pressure, using powers of 80 W and 180 W for Ni and Mo, respectively, onto the following substrates: Ni foam, Ni foil, Carbon cloth (C–cloth), indium tin oxide (ITO) coated glass and fluorine doped tin oxide (FTO) coated glass. The substrate holder was rotated at 3 rpm to improve homogeneity of the films. The film thickness was determined by a Veeco Dektak 150 surface profilometer to 160±10 nm.

Ni-Fe based catalysts were prepared on Ni foam by a hydrothermal method. Ni foam (about 2.5 cm x 3 cm) was reduced in 3 M HCl using an ultrasonic cleaner for 35 min, then washed in deionized water and ethanol for 6 min, and blown dry by nitrogen. Meanwhile, 0.29 g ($Ni(NO_3)_2$•$6H_2O$), 0.4 g $Fe(NO_3)_3$•$9H_2O$ and 0.6 g $CO(NH_2)_2$ were dissolved in 80 ml deionized water under magnetic stirring for 30 min to ensure homogeneity. After dissolution, the aqueous solution was transferred into a 100 ml Teflon-lined stainless steel autoclave. The Ni foam was vertically immersed into the solution with the topside protected by a Kapton tape. Film growth took place for 12 h at 120 °C in the autoclave. After cooling down, the sample was ultrasonically washed by deionized water and ethanol in sequence for 6 min to remove residual reactants and then dried at 80 °C for 12 h in air.

Electrochemical measurements were carried out in a three-electrode cell at room temperature. The sample was used as working electrode, while Ag/AgCl and a Pt wire or wire mesh were used as reference and counter electrodes, respectively. The electrolyte was 1M NaOH or 1M KOH (pH=14). The measured potential *vs.* Ag/AgCl was converted into the potential scale relative to the reversible hydrogen electrode (RHE) at pH=14. Impedance spectroscopy measurements were performed using a Zahner Zennium electrochemical workstation in the frequency range between 0.7 Hz and 100 kHz. The measurements were done using a 10 mV amplitude AC potential at applied DC potentials between 0 mV and –400 mV *vs.* RHE. Low/high negative values *vs.* RHE are denoted as low/high overpotentials for the HER. Analysis of the impedance data was carried out by the ZView software (Scribner Associates, Inc.).

3. Modelling of impedance spectra

A detailed theory for the frequency-dependent impedance of the HER, taking into account all the reaction steps (1)-(3) above, was derived by Harrington and Conway [12]. Their equations can be represented by either of four different equivalent circuits of which the two most used ones are depicted in Figure 1a and 1b. In addition to the elements describing the reaction, these circuits also contain the electrochemical double layer capacitance, C_{dl}, and a series resistance R_s due to the substrate and the electrolyte. It is observed that one of the circuits contains an inductance in the reaction part and the other a capacitance. However, if one allows L_a and C_a to take both positive and negative values, the two circuits are equivalent and represent the complete reaction kinetics. Harrington and Conway recommend the circuit in Figure 1a, since it allows a clear physical interpretation. In this case, the resistance R_{ct} is the reaction charge transfer resistance, while the L_a-R_{a1} series combination represents the additional part of the impedance response that is due to a changing surface coverage of adsorbed species [12]. Diard et al. performed a detailed analysis of the Volmer-Heyrovsky and Volmer-Tafel reaction sequences [13] and showed that the Volmer-Tafel case always leads to capacitive behaviour, while the Volmer-Heyrovsky case can exhibit either inductive or capacitive behaviour. It should be noted that these circuits represent ideal reaction kinetics and that in real cases it is often necessary to replace the capacitive elements by constant phase elements (CPE's).

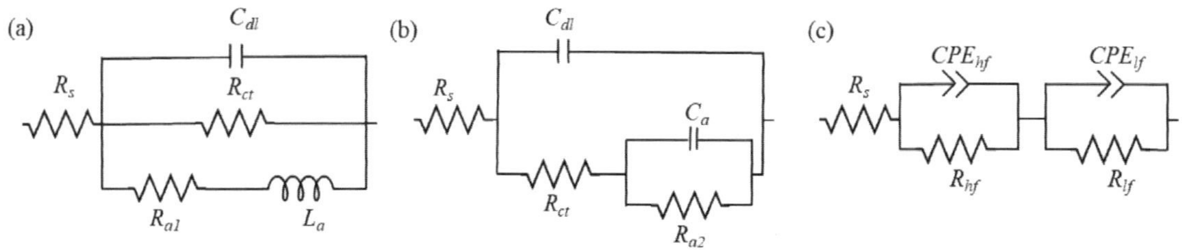

Figure 1. Equivalent circuits describing the (a) inductive behaviour and (b) capacitive behaviour predicted for the HER. (c) An empirical equivalent circuit often used to describe the impedance response of the HER.

Figure 1c depicts an empirical circuit that is frequently used to model water splitting reactions. It consists of the series resistance and two *R-CPE* parallel combinations denoted *hf* (high frequency) and *lf* (low frequency). Since ideal capacitive behaviour is seldom seen in impedance spectra, as noted above, we need CPE's instead of capacitances. The CPE behaviour [14] is believed to be related to sample inhomogeneities such as roughness and porosity of the electrodes. The circuit in figure 1c can be motivated by different arguments. Firstly, if the CPE's are replaced by ideal capacitances, it becomes equivalent to the circuit in Figure 1b. These two circuits will still give very similar results when we use CPE's instead of capacitances in both cases. Hence figure 1c may be an alternative way to represent the impedance of the HER. Secondly, in certain cases, the adsorptive part of the impedance related to surface coverage may be neglected in comparison with the charge transfer resistance. In this case the first *R-CPE* parallel combination in fig. 1c is taken to be due to the properties of a resistive interlayer on the electrode, while the second *R-CPE* combination gives CPE_{dl} and R_{ct}. It should be noted that a resistive interlayer is to be expected for the materials of interest in this paper. The transition metal based electrodes will oxidize in ambient air, immersion in an alkaline electrolyte will lead to growth of oxide/hydroxide compounds on the surface, and in addition, resistive barriers may occur at the interface with the substrate.

4. Results

In this section, we present preliminary results from an impedance spectroscopy study of Ni-Mo and Ni-Fe based catalysts and discuss some implications of the results. We first consider substrate effects on the impedance response of Ni-Mo and then discuss some aspects of a Ni-Fe catalyst showing very promising properties for water splitting.

4.1. Substrate effects for a Ni-Mo based catalyst

Ni-Mo was deposited onto different substrates with sheet resistances ranging from about 0.1 Ω for Ni foam to 43 Ω for ITO. Figure 2 shows experimental data in the complex impedance plane for a Ni-Mo based catalyst on (a) Ni foam and (b) ITO. Intermediate results were obtained for the other substrates. The data in figure 2a shows a well-developed semicircular arc and is typical of low resistance substrates. However, it was found that the data could not be accurately fitted by a single parallel R-CPE combination and instead the circuit in figure 1c had to be used. In figure 2b two distinct semicircular arcs are visible, and it is obvious that the circuit in figure 1c is again valid here. However, experimental data for all substrate cases could also be fitted with the circuit in figure 1b, by replacing the capacitances with CPE's. In order to distinguish these circuit models, we have to study the potential dependence of the fitting parameters.

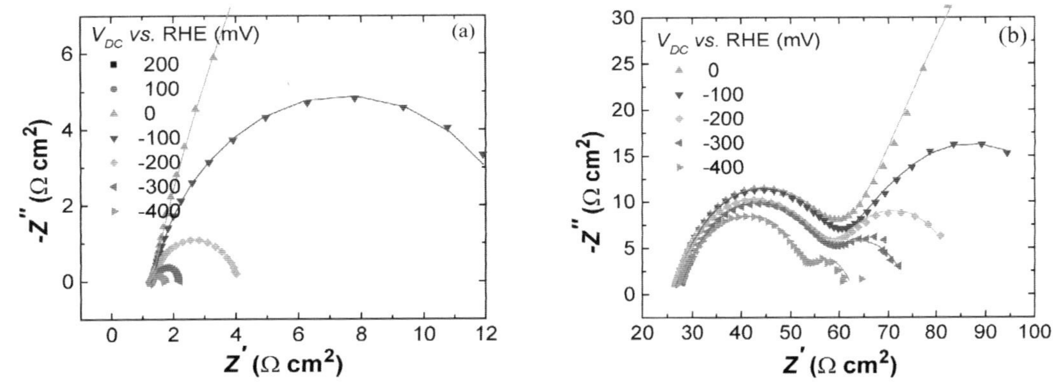

Figure 2. Imaginary *vs.* real impedance for a Ni-Mo based catalyst on (a) Ni foam and (b) ITO substrates, for different overpotentials *vs.* RHE. Experimental data are displayed by symbols and fits to the circuit of figure 1c are shown as lines.

The series resistance R_s was not significantly dependent on potential but showed a clear dependence on the sheet resistance of the substrate. It probably has contributions also from the electrolyte since it is slightly above 1 Ω for even the most conducting substrates. The CPE's display a more erratic behavior with no systematic dependence on potential. Figure 3 shows the potential dependence of the low (a) and high (b) frequency resistances obtained from fits to the circuit in figure 1c. The high-frequency resistance, R_{hf}, displays no systematic dependence on potential, but shows a strong dependence on the substrate resistance, and exhibits values of the order of, or even lower than R_s. On the other hand, R_{lf} shows a systematic dependence on potential, which is close to exponential at high negative potentials. Clearly, it must be identified with the reaction charge transfer resistance. However, in the fit to the circuit in figure 1b, it is R_a, which exhibits a systematic potential dependence while R_{ct} has values close to R_{hf}. Since the charge transfer resistance should depend on potential, the model behind figure 1b cannot describe all features of the experimental data, and we are obliged to choose the circuit in figure 1c for the analysis. Hence, we must identify R_{lf} with the reaction resistance and CPE_{lf} with the double layer CPE. The high-frequency elements cannot be due to the HER and the strong substrate dependence point to a resistive substrate-film interlayer, perhaps in combination with an oxide/hydroxide layer on the surface.

The reaction charge transfer resistance is closely related to the efficiency of the HER. It is seen in

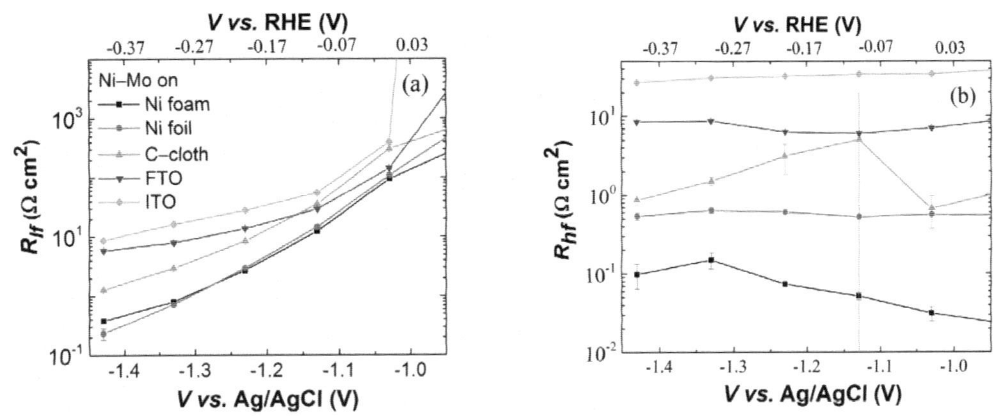

Figure 3. Low-frequency resistance (a) and high-frequency resistance (b) obtained from fits of the impedance data in figure 2 to the equivalent circuit in figure 1c. The potential is given both vs. Ag/AgCl (sat) and vs. RHE.

figure 3a that this quantity is lowest for the low resistance substrates, Ni foam and Ni foil, and much higher for the conducting oxide substrates. We conclude that a substrate with low sheet resistance is advantageous for the HER.

4.2. Impedance response of a Ni-Fe based catalyst

For this study, we chose Ni foam as the substrate since it gave the best results for the Ni-Mo based catalyst in sect. 4.1. Selected impedance data for the Ni-Fe based catalyst are shown in Figure 4 for two different overpotentials in the HER region. At low overpotentials, figure 4a, an asymmetric semicircular arc is found. On the other hand, at high overpotentials, an inductive behavior is clearly seen at low frequencies, where -Z'' crosses zero. This strongly indicates a Volmer-Heyrovsky mechanism, as discussed above. The crossover between capacitive and inductive behavior is also in qualitative agreement with the theoretical model of the Volmer-Heyrovsky reaction path [13]. The inductive behavior is modelled by the equivalent circuit in figure 1a, resulting in an excellent fit, as seen in figure 4b. However, neither of the theoretical models in figure 1a and 1b give satisfactory fits to the data in figure 4a. Another contribution to the semicircular arc is apparent, and the empirical circuit in figure 1c provides a good fit for the case of low overpotentials. In this case, we assume, as in

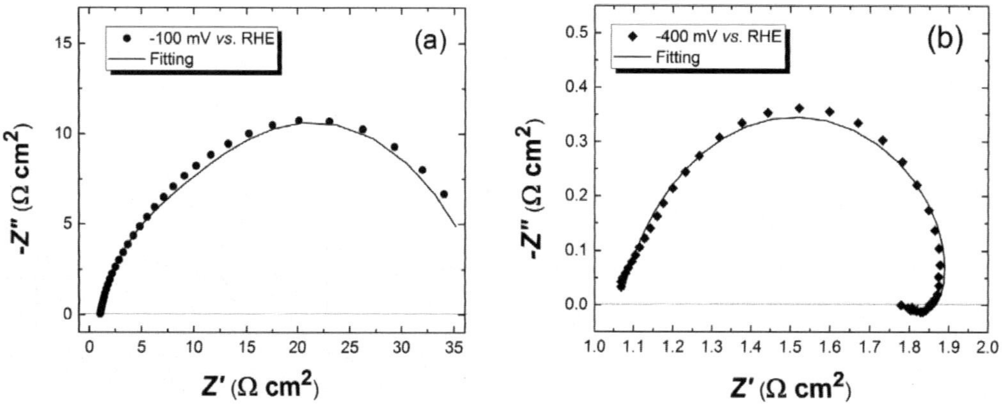

Figure 4. Imaginary *vs.* real impedance for a Ni-Fe based catalyst on Ni foam substrate, for two different overpotentials *vs.* RHE. Experimental data are displayed by symbols and fits to the circuit of figure 1c (a) and figure 1a (b) are shown as lines.

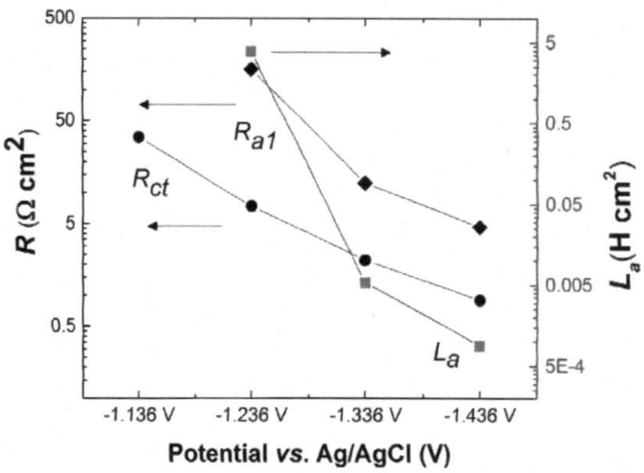

Figure 5. Charge transfer resistance $R_{ct,}$ adsorption resistance R_{a1} and inductance L_a are shown as a function of applied potential *vs.* Ag/AgCl (3M KCl). The parameters were obtained from fits of impedance data such as those in figure 4 to the equivalent circuit in figure 1a (circuit 1b was used for -1.136 V).

sec. 4.1, that the high-frequency parameters are due to a resistive interlayer on the substrate or the electrode while the high-frequency resistance is identified with the reaction charge transfer resistance.

We now discuss the potential dependence of the fit parameters. The inductance, L_a, as well as the resistance related to the adsorbed species, R_{a1}, depend strongly, and in a similar manner, on overpotential, as seen in figure 5. The charge transfer resistance is also potential dependent and exhibits very low values at high overpotentials. On the other hand, the series resistance and the double layer CPE do not significantly depend on potential, as is expected since these quantities depend primarily on the structure of the electrode and the electrochemical cell. The Ni-Fe based catalyst exhibits low reaction resistances both for the HER and the OER and is a very promising material that can be used for the complete water splitting process [15].

5. Conclusions

In this paper, we have explored the impedance response of transition metal based catalysts for the hydrogen evolution reaction. By equivalent circuit analysis, we obtain information on parameters describing the reaction kinetics as well as parameters dependent on the structure of the electrode and the electrochemical cell. The reaction kinetics is strongly dependent on the substrate used for the transition metal film, and it was shown that a substrate with low sheet resistance is to be preferred for the HER. In many cases, an impedance response in series with the HER impedance, possibly due to the film-substrate interlayer could be identified. The reaction charge transfer resistance was identified by its characteristic dependence on overpotential.

A Ni-Fe catalyst of prime interest for the OER was shown to display low reaction resistances for the HER as well. The inductive behaviour at high overpotentials indicates that the reaction proceeds according to the Volmer-Heyrovsky scheme. This catalyst is very promising for use on both the anode and cathode side of a water splitting device [15].

Acknowledgments

We gratefully acknowledge financial support from the Swedish Research Council (VR-2016-03713 and VR-2015-03814) and from the Fuel Cells and Hydrogen 2 Joint Undertaking (Grant No 735218), supported by the European Union (Horizon 2020), Hydrogen Europe and N.ERGHY.

References

[1] Edwards P P, Kuznetsov V L, Davis W I F and Brandon N P 2008 *Energy Pol.* **36** 4356-4362
[2] Seh Z W, Kibsgaard J, Dickens C F, Chorkendorff I, Nørskov J K and Jaramillo T F *Science* 2017 **335** eead4998/1-eead4998/12
[3] Roger I, Shipman M A and Symes M D 2017 *Nat. Rev. Chem.* **1** 0003/1-0003/10
[4] Shetty S, Sadiq M M J, Bhat D K and Hegde A C 2017 *J. Electroanal. Chem.* **796** 57-65
[5] Bockris J O'M and Potter E C 1952 *J. Electrochem. Soc.* **99** 169-186
[6] Lasia A 2002 Applications of the Electrochemical Impedance Spectroscopy to Hydrogen Adsorption, Evolution and Absorption into Metals *Modern Aspects of Electrochemistry* vol. 35 ed Conway B E and White R E (New York: Kluwer/Plenum) pp. 1-49
[7] Vilekar S A, Fishtik, I and Datta R 2010 *J. Electrochem. Soc.* **157** B1040-B1050
[8] Bockris J O'M 1956 *J. Chem. Phys.* **24** 817-827
[9] Doyle R L and Lyons M E G 2013 *Phys. Chem. Chem. Phys.* **15** 5224-5237
[10] Shinagawa T, Garcia-Esparza A T and Takanabe K 2015 *Sci. Rep.* **5** 13801/1-13801/21
[11] Diard J-P, Landaud P, Le Gorrec B and Montella C M 1988 *J. Electroanal. Chem.* **255** 1-20
[12] Harrington D A and Conway B E 1987 *Electrochim. Acta* **32** 1703-1712
[13] Diard J-P, Le Gorrec B and Maximovitch S 1990 *Electrochim. Acta* **35** 1099-1108
[14] Brug G J, van den Eeden A L G, Sluyters-Rehbach M and Sluyters J H 1984 *J. Electroanal. Chem.* **176** 275-295
[15] Qiu Z, Tai C-W, Niklasson G A and Edvinsson T, to be published.

IOP Conf. Series: Materials Science and Engineering **503** (2019) 012006 doi:10.1088/1757-899X/503/1/012006

TiO_2 thin films by ultrasonic spray pyrolysis

Z Chen, I Dündar, I Oja Acik and A Mere

Laboratory of Thin Film Chemical Technologies
Department of Materials and Environmental Technology
Tallinn University of Technology
Ehitajate tee 5, EE-19086 Tallinn, Estonia
E-mail: ilona.oja@ttu.ee

Abstract. Titanium dioxide (TiO_2) thin film have attracted wide interest as photovoltaic, dielectric and photocatalytic material. In this study, the ultrasonic spray pyrolysis method was used to fabricate TiO_2 thin films because of its low-cost operation, absence of vacuum system and convenience of use. The films were sprayed from solution containing titanium(IV)isopropoxide, acetylacetone and ethanol onto microscopy glass and n-type Si(100) substrates at temperatures of 200 to 500 °C. The resulting films were annealed at 500 °C and 700 °C for 1 hour in air. In the present study, optical, structural and electrical properties of produced TiO_2 thin films were investigated as a function of deposition temperature. TiO_2 films deposited ≤ 400 °C show optical transmittance of ca. 80% in the visible region. The band gap of TiO_2 films decreased from 3.52 to 3.25 eV as the deposition temperature increased from 200 to 500 °C. As-deposited films prepared below 500 °C were amorphous, whereas crystalline anatase films were obtained at 500 °C. Further annealing at 700 °C in air led to anatase crystalline formation when films were deposited below 500 °C whereas the films deposited at 500 °C consist of anatase and rutile phases.

1. Introduction

Titanium oxide (TiO_2) is one of the most important semiconductor oxides. Since the first report by Fujishima and Honda (1972) on the photolysis of water by TiO_2 [1], numerous studies in the properties of nanocrystalline TiO_2 have been generated due to their promising applications in photocatalysis [2], photoelectrolysis [3], dye-sensitized solar cells [4] and gas sensors [5], etc.

TiO_2 belongs to the family of transition metal oxides. It is known that TiO_2 has three polymorph phases: anatase, brookite, rutile [6]. Anatase and rutile have a crystalline structure that corresponds to the tetragonal system while brookite has an orthorhombic crystalline structure [7]. Rutile phase is thermodynamically stable at high temperatures. The anatase phase is metastable and transforms irreversibly to rutile at elevated temperatures [8]. The brookite phase is the rarest of the natural TiO_2 polymorphs and is the most difficult phase to prepare in the laboratory [9]. Anatase phase has a band gap of 3.2 eV, while the rutile phase has a smaller band gap of 3.0 eV [10].

In general, anatase phase is preferred in solar cells and photocatalytic applications because anatase phase has a larger band gap, potentially higher conduction band edge energy and lower electron-hole pair recombination rate [11].

Various methods have been used for preparing nanocrystalline titania thin films, such as sol-gel [12], screen-printing [13], dip-coating [14], chemical vapour deposition [15] and ultrasonic spray pyrolysis [16]. Among all these methods, ultrasonic spray pyrolysis method has been drawn considerable attention due to its simplicity, cost-efficiency and convenience for fabricating TiO_2 thin films. In this paper, TiO_2 thin film was deposited on the silicon and microscopy glass substrates by using ultrasonic spray pyrolysis method. The aim of this work is to deposit TiO_2 thin films at different substrate temperature and to investigate its structural, optical and electrical properties.

2. Experimental

TiO_2 thin films were deposited onto microscopy glass and c-Si substrates at different substrate temperature using ultrasonic spray pyrolysis method. The precursor solution is composed of titanium(IV) isopropoxide (TTIP) as a titanium source, acetylacetone (AcAc) as a stabilizer and ethanol as solvent. 0.2molL^{-1} TTIP concentration and TTIP:AcAc molar ratio of 1:4 were used as starting solution. The solution was atomized by a ultrasonic generator of 1.7MHz frequency, produced aerosol was carried directly to the heated substrates using compressed air as carrier gas in a flow rate of 5 L/min. The deposition temperature (Ts) vary in the range 200 to 500 °C. The number of spray cycles were optimized to six.

The as-deposited films grown on glass and Si-substrates were annealed for 1 hour at 500 °C in air. The films on silicon substrates where additionally thermally treated at 700 °C for 1 hour in a laboratory furnace. The structural property of the samples was investigated by X-ray diffraction (XRD) method. XRD patterns were recorded by a Rigaku Ultima IV diffractometer with Cu Kα radiation (λ=1.5406 Å, 40 kV at 40 mA) using Si strip detector. The measurements were performed in 2 theta/theta configurations with the scan range of 20-60 deg, with a step of 0.02 deg and a scanning speed of 2 deg min^{-1}. The mean crystallite size was calculated by the Scherrer method from the FWHM (full width at half maximum) of the (101) reflection of TiO_2 anatase phase. The surface morphology of the samples was investigated by Scanning Electron Microscopy (SEM) Zeiss HR FESEM Ultra 55 measurements with an acceleration voltage of 4.0 kV. The total transmittance of TiO_2 thin films is measured by Jasco V-670 spectrophotometer with a spectral range between 250 nm and 1200 nm. The film thickness was calculated by using interference fringes from the optical spectrum, and refractive index published in our previous papers [17]. Current-Voltage (I-V) measurement, taken from -1V to +1V, was performed to investigate the electrical properties of TiO_2 thin film on n-Si substrate. The I-V measurement was performed using graphite contact and Si as a second contact. The Autolab PGSTAT 30 system was used for measuring of I-V curves of the films.

3. Results and discussion

3.1. Morphology

The SEM images of the TiO_2 films deposited onto glass substrate at deposition temperatures of 200 and 500 °C and annealed at 500 °C for 1h in air are presented in Figure 1. As seen, the TiO_2 films are smooth, dense and free from cracks. The TiO_2 film deposited at 200 °C has a plane surface structure. The films deposited at 500 °C have larger well-distinguished grains with a size of ca. 50 nm.

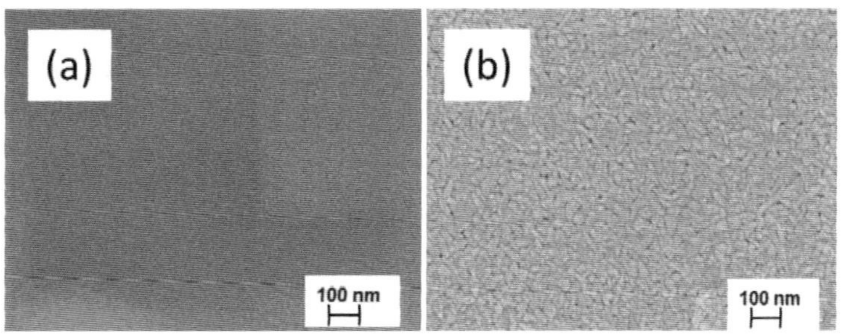

Figure 1. SEM surface images of TiO_2 thin films deposited on glass at (a) 200 °C, (b) 500 °C, and annealed at 500 °C for 1 h.

3.2. Structural properties

The XRD patterns of TiO_2 thin films on glass substrates deposited at different deposition temperatures from 200 to 500 °C are given in Figure. 2. The diffraction peaks at 25.3, 37.7, 48.1, 53.9 and 55.1deg are found to correspond to the (101), (004), (200), (105) and (211) anatase planes [18].

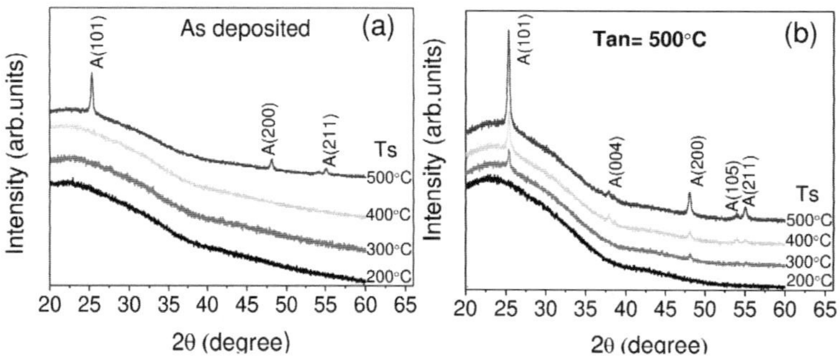

Figure 2. XRD patterns of TiO_2 thin films on glass substrates deposited at various temperatures from 200 to 500 °C: (a) as-deposited thin films, (b) thin films annealed at 500 °C.

XRD studies showed that the as-deposited films prepared at substrate temperatures below 500 °C are amorphous. The anatase peak (101) becomes apparent for the films deposited at 500 °C. Annealing at 500 °C results in anatase phase for films deposited at or above 300 °C.

Figure 3 shows XRD patterns of TiO_2 thin films deposited on silicon substrates. XRD result showed that the films deposited at or below 300 °C are amorphous. However, the films deposited at or above 400 °C have anatase structure. Annealing at 500 °C results in the formation of anatase phase regardless of the deposition temperature. Further annealing at 700 °C has significantly changed the film structure, the films deposited below or at 400 °C remain anatase, however, the films deposited at 500 °C composed of a mixture of anatase and rutile phases.

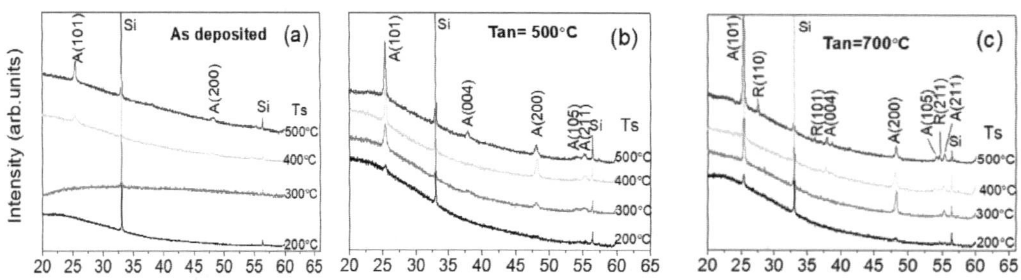

Figure 3. XRD patterns of TiO_2 thin films on silicon substrates. (a) as-deposited thin films, (b) thin films annealed at 500 °C, (c) thin films annealed at 700 °C.

The mean crystallite size of TiO_2 thin films was calculated from the (101) peak of anatase phase by the Scherrer formula. The mean crystallite size of anatase TiO_2 films annealed at 500 °C increased from 20 nm to 32 nm with increasing the deposition temperature from 200 to 500 °C, irrespective of the substrate.

3.3. *Optical properties*

The optical transmittance spectra of TiO_2 thin films deposited onto glass substrate at temperatures from 200 to 500 °C and followed by annealing at 500 °C are presented in Figure 4.

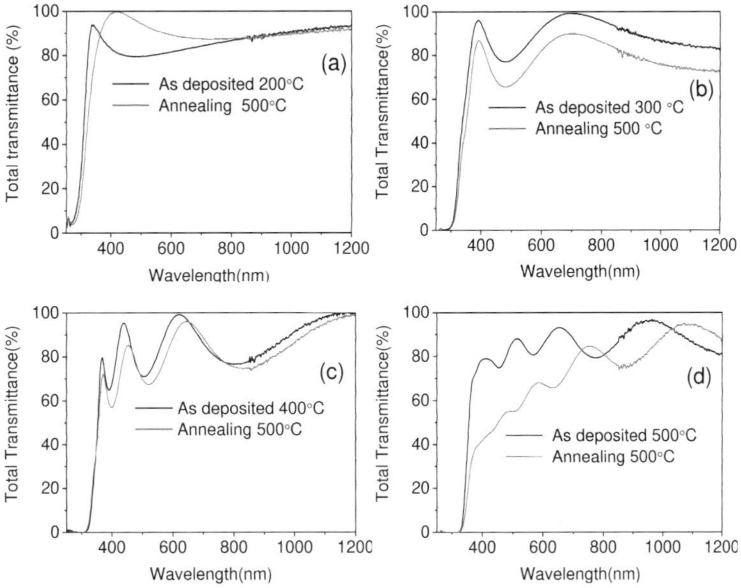

Figure 4. Total transmittance spectra of TiO_2 thin films deposited at various temperatures : a) 200 °C, b) 300 °C, c) 400 °C, d) 500 °C on glass substrates and followed by annealing at 500 °C.

The TiO_2 thin films were highly transparent in the visible region with total transmittance of 80 %. The total transmittance decreased with increasing the deposition temperature. The interference patterns indicate the homogeneity of the film. The decrease in the optical transmittance after annealing can be related to the increase of the light diffusion with the crystallite size and particle aggregation [19].

The optical band gap was determined using the Tauc expression [20],

$$\alpha(hv) = B(hv - E_g)^m \tag{1}$$

where B is a proportionality constant, hv is the photon energy, Eg is the band gap and m=2 for indirect optical transitions. The optical band gap was obtained by extrapolating the linear part of the plot of $(\alpha hv)^2$ against hv to the photon energy axis by assuming direct band gap is shown in Figure 5. The band gaps of the as-deposited films were 3.52, 3.34,3.27, 3.25 eV, when deposited at 200, 300, 400 and 500 °C, respectively. The band gap of TiO_2 films decreases slightly after annealing at 500 °C and is found to be 3.51, 3.35, 3.27 and 3.12 eV, respectively. The decrease in the optical band gap of the TiO_2 films with annealing temperature might be the result of their increased crystallinity [21] and also their increase in the mean crystallite size.

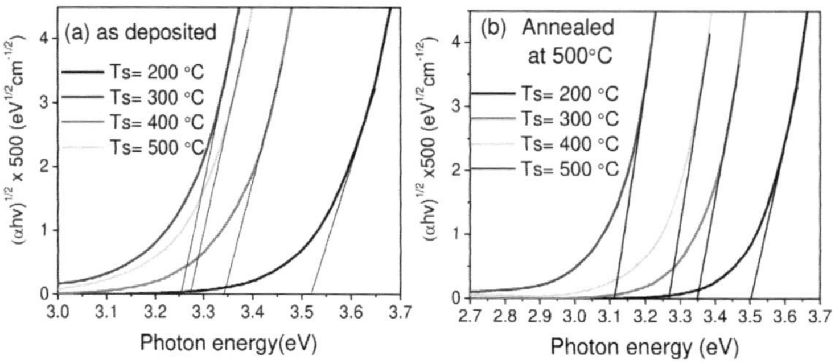

Figure 5. The optical band gap of TiO_2 thin films on glass substrates: (a) as-deposited films deposited at substrate temperature from 200 to 500 °C (b) as-deposited films followed by annealing at 500 °C.

The thickness of as-deposited films increased from 110 to 620 nm with increasing deposition temperature from 200 to 500 °C. The annealing at 500 °C reduced the film thickness, remaining in the range of 80 to 500 nm, with increasing the deposition temperature from 200 to 500 °C. Since the solution feed rate was constant, then it is implicit that the film thickness increased with increasing deposition temperature owing to the corresponding increase in the reaction rate. The film thickness decreased after annealing at 500 °C, compared to as-deposited films, is related to the burning out of the organic residues in the film [22].

3.4. *Electrical resistivity*

Figure 6 shows the resistivity of the as-deposited and annealed TiO_2 films on silicon substrates. The resistivity of the films decreases with increasing deposition temperature from 200 to 300 °C, which may be due to the change of crystal structure from amorphous to crystalline anatase phase. The resistivity also decreased when increasing the annealing temperature; this phenomenon is due to an increase in the grain size which leads to a decrease in the grain boundaries and hence resistivity [23].

The equation used for the calculation of resistivity is as follows:

$$\rho = \frac{V}{I} \cdot \frac{S}{T} \tag{2}$$

Where, ρ - resistivity, Ω·cm; V-Voltage, V; I- current, A; T- TiO_2 film thickness, cm; S, the surface area of contact, cm^2.

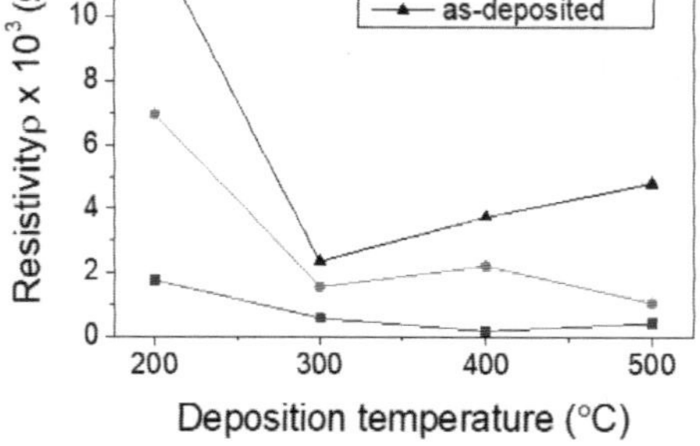

Figure 6. Resistivity of TiO_2 films deposited on silicon substrates at various substrate temperatures (T_s) and after various annealing temperatures.

4. Conclusions

Ultrasonic spray pyrolysis method was used to deposit TiO_2 film and the effect of deposition and annealing temperature on the structural, optical, electrical properties was investigated. The XRD data revealed that the as-deposited films grown below 400 °C onto glass substrate and below 300 °C onto Si substrate are amorphous. Annealing at 500 °C results in the formation of anatase structure, if the films are deposited at temperatures ≥ 300 °C onto glass substrates or in the temperature range 200-500 °C onto Si substrates. The mean crystallite size of anatase structure remains in the range of 20-32 nm irrespective of the deposition temperature and substrate. Further annealing of TiO_2 films on Si substrate at 700 °C led to the formation of the mixture of anatase and rutile phase if deposited at 500 °C and increase in the average crystallite size of the anatase structure to the range of 30-40 nm. As-deposited TiO_2 films show high optical transmittance and band gap values in the range of 3.52-3.25 eV if deposited at 200-500 °C. Film thickness was found to increase with deposition temperature and decrease after annealing at 500 °C. The resistivity of TiO_2 films depended on the deposition temperature and was found to decrease with increasing the annealing temperature.

Acknowledgement

This study was financially supported by the Estonian Ministry of Education and Research project IUT19-4, and by the European Union through the European Regional Development Fund project TK141 "Advanced materials and high-technology devices for energy recuperation systems". The authors acknowledge Dr. Valdek Mikli for SEM images.

References

[1] Fujishima A and Honda K 1972 *Nature* **238** 37
[2] Paul S, Choudhury A and Bojja S 2013 *Micro. Nano. Lett* **8** 184
[3] Shin K, Yoo J B and Park J H 2013 *J. Power. Sour* **225** 263
[4] Gong J W, Sumathy K, Qiao Q Q and Zhou Z P 2017 *Renew. Sust. Energ. Rev* **68** 234
[5] Bai J and Zhou B X 2014 *Chem. Rev* **114** 10131
[6] Reyes-Coronado D, Rodriguez-Gattorno G, Espinosa-Pesqueira, M E, Cab C, de Coss R and Oskam G 2008 *Nanotechnology* **19**
[7] Nouri E, Mohammadi M R, Xu Z X, Dracopoulos V and Lianos P 2018 *Phys. Chem. Chem. Phys* **20** 2388
[8] Lira-Cantu M, Chafiq A, Faissat J, Gonzalez-Valls I and Yu Y H 2011 *Sol. Energy Mater. Sol. Cells* **95** 1362
[9] Mahalingam S and Edirisinghe M J 2007 *Appl. Phys. A. Mater. Sci. Process* **89** 987
[10] Paul S and Choudhury A 2014 *Appl. Nanosci* **4** 839
[11] Luttrell T, Halpegamage S, Tao J G, Kramer A, Sutter E and Batzill M 2014 *Sci. Rep* **4** 4043
[12] Yu J G, Zhao X J and Zhao Q N 2000 *Thin Solid Films* **379** 7
[13] Sauvage F, Chen D H, Comte P, Huang F Z, Heiniger L P, Cheng Y B, Caruso R A and Graetzel M 2010 *ACS. Nano* **4** 4420
[14] Negishi N and Takeuchi K 2001 *J. Solgel. Sci. Technol* 2001 **22** 23
[15] Sun H F, Wang C Y, Pang S H, Li X P, Tao Y, Tang H J and Liu M 2008 *J. Non. Cryst. Solids* **354** 1440
[16] Ennaceri H, Boujnah M, Taleb A, Khaldoun A, Saez-Araoz R, Ennaoui A, El Kenz A and Benyoussef A 2017 *Int. J. Hydrogen. Energy* **42** 19467
[17] Oja I, Mere A, Krunks M, Nisumaa R, Solterbeck C H and Es-Souni M 2006 *Thin. Solid. Films* **515** 674

[18] Fazli F I M, Ahmad M K, Soon C F, Nafarizal N, Suriani A B, Mohamed A, Mamat M H, Malek M F, Shimomura M and Murakami K 2017 *Optik* **140** 1063
[19] Ben Karoui M, Kaddachi Z and Gharbi R 2015 *Proc. Tunisia-Japan Symposium - R&D of Energy and Material Sciences for Sustainable Society* (Gammarth, 2014) *596*
[20] Tauc J 1968 *Mater Res Bull* **3** 37
[21] Lin C P, Chen H, Nakaruk A, Koshy P and Sorrell C C 2013 *Proc. 10th Eco-Energy and Materials Science and Engineering Symposium* (Ubon Ratchathani, 2012) **34** 627
[22] Oja I, Mere A, Krunks M, Solterbeck CH and Es-Souni M 2004 *Proc. Symposium on Functional Nanomaterials for Optoelectronics and other Applications* (Warsaw, 2003) **99** 259
[23] Bakri A S, Sandan M Z, Adriyanto F, Raship N A, Said N D M, Abdullah S A and Rahim M S 2017 *Proc. Int. Conf. on Engineering, Science and Nanotechnology* (Solo, 2016) 1788

IOP Conf. Series: Materials Science and Engineering **503** (2019) 012007 doi:10.1088/1757-899X/503/1/012007

Deuterium concentration's in austenitic stainless steel by deuterium irradiation. Effects temperature irradiation

Oleksandr Morozov, Volodymyr Zhurba, Oleksandr Mats, Viktoria Progolaieva, Viktoriia Seliukova and Nikolay Kochnev

NSC "Kharkiv Institute of Physics and Technology" Kharkiv, Ukraine

E-mail: morozov@kipt.kharkov.ua

Abstract. Using the deuterium thermodesorption spectra from the 18Cr10NiTi steel samples exposed to different doses, we have plotted the total amount of deuterium desorbed from the sample as a function of the radiation dose and the irradiation temperatures. *Temperature 100 K.* The maximum attainable concentration of deuterium in steel is C=1 (at.D/at.met.=1/1). At C≥0.5, two hydride phases are formed in the steel, the decay temperatures of which are 240 K and 275 K. The hydride phases are formed in the bcc structure resulting from the martensitic structural transformation in steel [1]. *Temperature 295 K.* The medium-dose region is characterized by radiation induced action on the steel in the presence of hydrogen. The process results in the formation of the energy-stable crystalline nanostructure of steel, having a developed network of intercrystalline boundaries. The basis for this developed network of intercrystalline boundaries is provided by the amorphous state. The total concentration of the accumulated deuterium in the region of medium implantation doses makes 7 to 8 at.% [2]. *Temperature 380; 420; 600 and 900 K.* In a deuterium thermodesorption spectra the extended area of deuterium desorption in a range of temperatures 450-900 K is observed, caused by formation of local structure in a radiation induced layer. Formation of local structure can be caused by a segregation of a steel component in the course of deuterium implantation (radiation induced local structure at deuterium presence).The total concentration of the accumulated deuterium in the region of medium implantation doses makes 1 to 3 at.%.

1. Introduction

The investigation into regularities of hydrogen interaction with metals and alloys over a wide range of temperatures and pressures still remains a currently central problem in material physics from both the scientific and practical standpoints. The hydrogen accumulation in structural and functional materials is an extremely hazardous phenomenon, which leads to hydrogen degradation of materials and to possible unforeseen equipment failures. The degradation of materials increases due to hydrogen interaction with the whole range of crystal structure imperfections of solids such as interstitial and substitutional impurities, vacancies and their complexes, dislocations and their pileups, subgrain and grain boundaries, and phase components. Stainless steel is one of the most useful classes of engineering materials. For example, austenitic steel is used for manufacturing vessel internals of fission-type reactors. A wide use of austenitic stainless steels as structural fission reactor elements calls for a detailed knowledge of their behavior under conditions of radiation influence, accumulation of gas impurities (hydrogen isotopes, above all). Hydrogenation of the face-centered cubic (fcc) iron-based alloys, which constitute an array of austenitic stainless steels, can cause phase transformations: fcc (γ)→bcc (α^*) and fcc (γ)→hcp (ε) [3][4]. The hydrogen-induced phases are sometimes considered as pseudohydrides. It is generally assumed that the role of hydrogen consists in creation of a particular

IOP Conf. Series: Materials Science and Engineering **503** (2019) 012007 doi:10.1088/1757-899X/503/1/012007

stress state that triggers the phase transformation. Morozov and co-workers [1] showed the steel surface layer exposed to deuterium ions at Tirr.= 100 K to doses above 8.2 $\times 10^{17}$ D/cm^2, exhibits austenite, α-martensite needles.

At room temperature, deuterium has a high diffusive mobility, the surface open to diffusion, and thereby is intensively desorbed from austenitic steel during its implantation. As is known, at room temperature, nearly all the defects formed in the course of irradiation, and above all, the Frenkel pairs are recombining [5]. In the process of deuterium ion implantation in the steel being at temperature 295 K, there occurs the formation of a great number of chaotically distributed dislocations, small precipitates, separate ordered segregations, which provide the basis for the occurrence of a nanocrystalline microstructure at a dose of 1.68×10^{18} D/cm^2 [2].

The present paper presents the results from studies the TDS deuterium of the sample irradiated to different doses, we have plotted the total concentration of deuterium, desorbed from the specimen as a function of radiation dose and temperature irradiation.

2. Experimental procedure

Austenitic steel 12Cr18Ni10Ti (0.12 wt.%C) and 08Cr18Ni10Ti (0.08 wt.%C) samples were investigated. The samples were homogenized at 1,350 K for 30 min in a vacuum of 5×10^{-5} Pa. Deuterium introduction into the samples was realized through implantation of 24 keV D_2^+ ions at a current density of 2 to 5 μA/cm^2, with doses ranging from 1×10^{15} to 4×10^{18} at.D/cm^2; the irradiation temperatures were T ≥ RT. The implanted samples were subjected to heating immediately after the ion beam switching-off at a rate of 3.5 – 5 K/s down to a temperature of 1700 K, with simultaneous registration of D_2^+ ions ion desorption spectrum (4 amu).

3. Results and discussion

Using the deuterium thermodesorption spectra from the 18Cr10NiTi steel samples exposed to different doses, we have plotted the total amount of deuterium C(F) desorbed from the sample as a function of the radiation dose F and the irradiation temperatures.

3.1 Deuterium Thermodesorption Spectra at Temperatures 380; 420; 600 and 900 K

The deuterium TD spectrum of steel 12Cr18Ni10Ti irradiated at 380 K (see Figure 1, curve b) exhibits only a low-intensity, wide temperature-range region of deuterium desorption, which begins practically from the irradiation temperature and extends up to ~ 1200 K. Here, it is clearly seen that this region is composed of three peaks with the maximum temperatures of 480, 700 and 1020 K. The presence of these peaks in the TD spectrum testifies that in the process of deuterium implantation in the steel 12Cr18Ni10Ti local structural compositions from steel components were formed. We have made this conclusion on the basis of the data obtained in ref. [32-33], where it was demonstrated that the ion implantation of nitrogen, oxygen, carbon, helium could lead to the formation of such local structures as nitrides, carbides, the presence of which in the deuterium TD spectra manifests itself in the form of

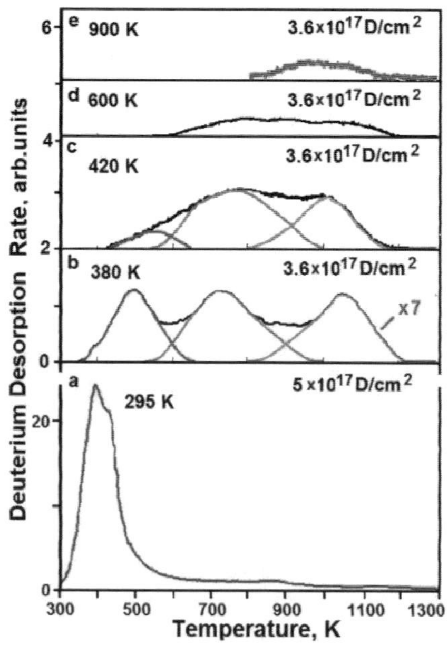

Figure 1. Deuterium thermal desorption spectra of steel 12Cr18Ni10Ti samples, irradiation with deuterium at temperatures of 295 K [1], 380, 420, 600, 900 K to a dose of 3.6×10^{17} D/cm^2

occurrence of additional temperature ranges of deuterium desorption (hydrogen diagnostics). With further increase in the irradiation temperature, both the amount of retained deuterium and the temperature range of deuterium desorption get reduced.

It is assumed that carbon may be one of the steel components capable of stimulating the local structure formation. This assumption is supported by the data on the carbon behavior in nickel, reported in ref [6]. It has been demonstrated there that on heating of nickel, which does not form chemical compounds with carbon, the latter diffuses to the crystallite boundaries. It is reasonable to expect that the carbon behavior in the austenitic stainless steel is similar to the behavior of carbon in nickel. On this basis it can be suggested that the irradiation of steel with deuterium might stimulate the formation of local structures from steel components and carbon.

3.2 Effect of carbon concentration in steel on temperature ranges of deuterium retenthion retention

For estimating the possibility of local structure formation from steel components and carbon, studies were made into deuterium TD spectra from steels having different carbon concentrations: 12Cr18Ni10Ti (0.12 wt.%C) and 08Cr18Ni10Ti (0.08 wt.%C). The deuterium TD spectra from these steels, irradiated at temperatures of 295 and 900 K, are presented in Figure 2. As may be seen from the figures, in the spectrum from the steel with a lower carbon concentration (0.08 wt. % C) there is no highest-temperature spectral region, i.e., there is no additional peak with the maximum temperature of 1020 K. At the same time, in the case of 12Cr18Ni10Ti steel irradiation with a higher carbon concentration (0.12 wt. %C), the local structure formation takes place, which manifests itself in the occurrence of the additional peak with the maximum temperature of 1020 K. Relying on the data obtained, we have concluded that the carbon present in the 12Cr18Ni10Ti steel stimulates the steel components segregation and the local structure formation.

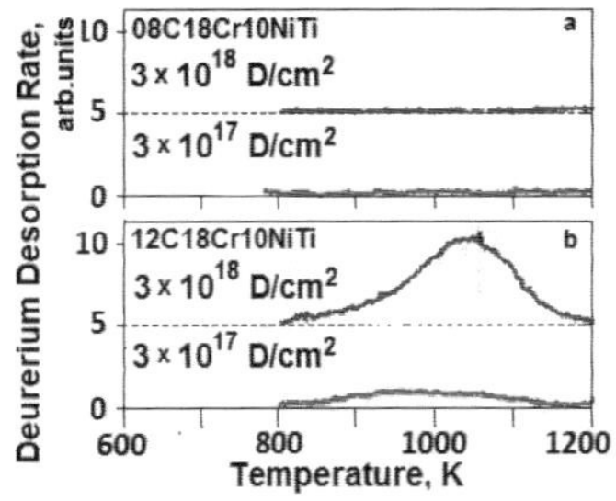

Figure 2. Deuterium thermal desorption spectra from steel samples: a) 08Cr18Ni10Ti, b) 12Cr18Ni10Ti, irradiation at Tirr=900 K

3.3 The amount of retained deuterium as a function of temperature and radiation dose

Using the deuterium thermodesorption spectra from the Cr18Ni10Ti steel samples exposed to different doses, we have plotted the total amount of deuterium C(F) desorbed from the sample as a function of the radiation dose F. This function is presented in Figures 3a and 3b. The total amount of the desorbed deuterium was determined from the area under the gas release curve.

The concentrations of implanted deuterium were estimated with consideration for both the quantity of metal atoms in the implantation layer and the amount of desorbed deuterium. At that, the implanted deuterium was assumed to have low diffusion mobility in the steel cooled down to Tirr=100 K, and practically all its amount be in the implantation layer. The calculation data have shown that the Cr18Ni10Ti steel saturation with deuterium is attained at the deuterium concentration C= 1.0 at.D/at.met. [1].

At Tirr= 295 K (RT), in the range of low implantation doses, the concentration of accumulated deuterium varies between 2.5 to 3 at.%. On completion of deuterium solid-solution formation in steel, the newly implanted deuterium atoms give impetus to further structural transformations, which are due to both the radiation action and the increasing concentration of implanted deuterium. In this case, the amount of retained deuterium increases at the expense of growing intensity of the high-temperature region of the spectrum with the peak temperatures 440 and 500 K. At a dose of 8×10^{17} D/cm^2, the accumulation of retained deuterium at the peak with Tm = 440 K practically fully ceases. The deuterium concentration in this phase state remains practically unchanged during further deuterium ion implantation, and is estimated to be 7–8 at.%. The irradiation temperature increase is accompanied by the reduction in the retained deuterium concentration down to ~ 1 at.% at temperatures between 420 to 600 K.

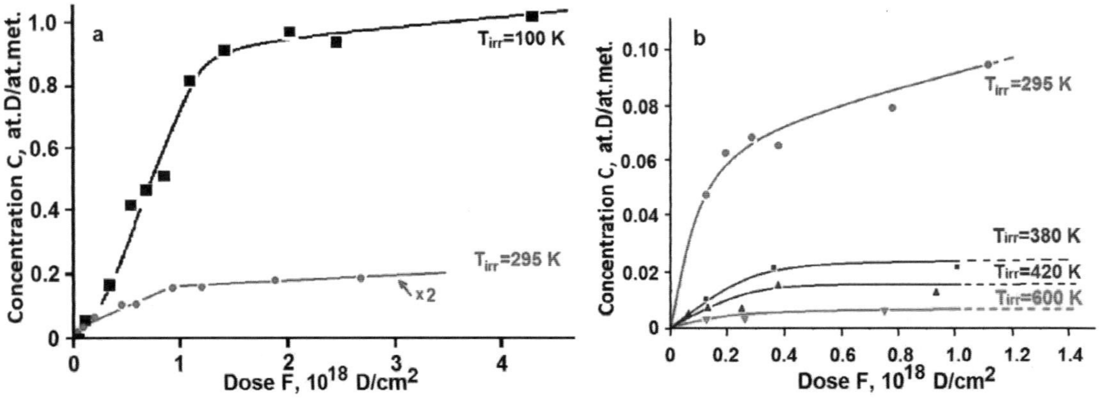

Figure 3. Desorbed deuterium concentration versus irradiation dose for the Cr18Ni10Ti steel implanted at different temperatures

4. Conclusions

Temperature 100 K. The maximum attainable concentration of deuterium in steel is C=1 (at.D/at.met.=1/1). At C≥0.5, two hydride phases are formed in the steel, the decay temperatures of which are 240 K and 275 K [1].

Temperature 295 K. The medium-dose region is characterized by radiation induced action on the steel in the presence of hydrogen. The process results in the formation of the energy-stable crystalline nanostructure of steel, having a developed network of intercrystalline boundaries. The basis for this developed network of intercrystalline boundaries is provided by the amorphous state. The total concentration of the accumulated deuterium in the region of medium implantation doses makes 7 to 8 at.% [2].

Temperature 380; 420; 600 and 900 K. In a deuterium thermodesorption spectra the extended area of deuterium desorption in a range of temperatures 450-900 K is observed, caused by formation of local structure in a radiation induced layer. Formation of local structure can be caused by a segregation of a steel component in the course of deuterium implantation (radiation induced local structure at deuterium presence).The total concentration of the accumulated deuterium in the region of medium implantation doses makes 1 to 3 at.%.

References

[1] Morozov O, Zhurba V, Neklyudov I, Mats O, Rud A, Chernyak N and Progolaieva V 2015 *Nanoscale Research Letters* **10** 154
[2] Morozov O, Zhurba V, Neklyudov I, Mats O, Progolaieva V and Boshko V 2016 *Nanoscale Research Letters* **11** 44
[3] Borchers C, Michler T and Pundt A. 2008 *Adv. Eng. Mater.* **10** 11–23
[4] Mine Y, Horita Z and Murakami Y 2009 *Acta Mater.* **5** 2993–3002
[5] Zelensky V F, Neklyudov I M and Chernyayeva T P. Radiation defects and swelling of metals (in Russian) 1988 *Naukova dumka* Kiev 230 p.
[6] Morozov O M and Neklyudov I M. 2014 *Proc. Internat. Conf. Nanomaterials: Applications and Properties* **1** 01NTF07 4 pp.

IOP Conf. Series: Materials Science and Engineering **503** (2019) 012008 doi:10.1088/1757-899X/503/1/012008

Preparation of ZnO modified TiO_2 nanoporous coatings and their photocatalytic properties

R Drunka, J Grabis, A Letlena, Dz Jankovica, A Krumina and I Steins

Riga Technical University, Faculty of Materials Science and Applied Chemistry, Institute of Inorganic Chemistry, Paula Valdena 7, Riga, LV1048

reinis.drunka@rtu.lv

Abstract. In the present work formation of active nanoporous TiO_2 photocatalysts and their modification with ZnO were studied. TiO_2 nanoporous coatings on Ti foil with surface area $100cm^2$ were prepared by using plasma electrolytic oxidation (PEO) method in 0.5M H_2SO_4 electrolyte and by using 160V DC voltage. For modification of TiO_2 coatings with ZnO spray pyrolysis method where used. After preparation samples were dried and calcined at 400°C for 2h in air. Photocatalytic properties of obtained samples were tested by using degradation method of methylene blue (MB) solution under UV and VIS light illumination. Degradation rate of MB were measured by using spectrophotometer. From obtained results reaction constant (k) in the pseudo first order reaction were calculated. As prepared ZnO modified TiO_2 nanoporous coating photocatalysts are with higher photocatalytic activity with respect to pure TiO_2 nanoporous coating photocatalyst.

1. Introduction

TiO_2 photocatalysis is widely used in a variety of applications and products in the environmental and energy fields. Various techniques had been widely applied to remove different types of toxic organic environmental pollutants from waste water by using methods of adsorption and degradation. One of the most promising and cost efficiently way to do it is in presence of photocatalysts and sunlight [1-2]. Due to its wide band gap (3.2eV) pure TiO_2 anatase as photocatalyst can be used only under UV irradiation. As only about 5% of sunlight that reaches the Earth contains UV irradiation modification of TiO_2 structure is required. By increasing the specific surface area and modification with various dopants efficiency of TiO_2 photocatalysts in VIS irradiation can be improved dramatically. As ZnO is well known photocatalyst with high activity in visible light as dopant it could improve light adsorption properties of TiO_2 and decrease the band gap of it. It can reduce the possibility of electron-hole recombination, causing efficient charge separation and higher photocatalytic reaction rates [3].
Plasma electrolyte oxidation is electro-chemical method that allows to obtain thin layer nanoporous oxide coatings in short time. TiO_2 coatings obtained on metal substrate have better mechanical properties and and it is easier to use in them continuously in photocatalysis because no filtration is required to remove them from the purified solution, as in the case of nanopowders [4].

Spray pyrolysis is a processing technique being considered in research to prepare thin films and ceramic coatings. It is a very simple and relatively cost-effective processing method that offers an extremely easy technique for preparing films of any composition. Typical spray pyrolysis equipment consists of an atomizer, precursor solution, substrate heater, and temperature controller [5].

The combination of both methods can allow to prepare on substrate surface fixed and visible light active photocatalysts in short time. Such materials can be used to solve environmental and energy problems.

2. Experimental

Coating of self-organized TiO_2 nanopore layer on Ti substrate was prepared by using plasma electrolytic oxidation of titanium foil 1 x 3cm (Sigma Aldrich, 0.25mm thick, 99.7%) in 0.5M H_2SO_4 electrolyte, using Pt foil as counter electrode. Before PEO process titanium foil was washed with acetone and etched in 1M HF for 30 sec than washed deionised water and dried in 110°C for 2 hours. Platinum foil was used as second electrode. The voltage was 160V DC and experimental process was 3 min long. During the PEO process micro sparks were noticed on the surface of titanium electrode. As obtained TiO_2 coating was washed with deionised water dried in 110°C for 2h and calcinated at 500°C for 2h in air.

Holmarc HO-HT-04BT spry pyrolysis system was used for modification of obtained TiO_2 nanoporous coatings with ZnO. 0.1M Zinc acetate solution was used as Zn source. Flow rate of solution atomization was 1ml per minute. The atomizer was programmed to move in XY coordinates in 140 x 140mm area with speed 400mm$\cdot$ sec^{-1} in X direction and 4mm$\cdot$ sec^{-1} in Y direction. Duration of spry pyrolysis process were in range of 0.5 to 10min. Obtained samples were calcinated at 400°C for 2 h in air.

Characterizations of physical properties as morphology, crystallization, chemical content were done by scanning electron microscopy (SEM; Tescan; Lyra), X-ray diffraction (XRD; Bruker D8 advance), X-ray fluorescence (XRF; Bruker Pioneer S4) analysis methods and instruments, respectively.

Photocatalytic activity of the prepared samples was evaluated by decomposition of MB water solution under visible light (Osram LEDVANCE floodlight, 100 W; 10000 lm). The MB solution (100 ml, $7.6\cdot10^{-3}g\cdot l^{-1}$) was introduced in a quartz reactor equipped with obtained ZnO modified porous TiO_2/Ti photocatalyst. The sources of light were at distance of 110 mm from the quartz beaker. During the experiments every 30 min MB sample (2 ml) was taken out for measurement. The degradation of MB under illumination in presence of the electrode was determined by measuring light absorption by solution at light wave length of 664 nm using JENWAY-6300 spectrophotometer. As reference with 100 % transmission was used deionized water.

3. Results and discussion

XRD patterns of obtained samples are shown in Figure 1. All samples showed strong maxima of titanium, anatase and rutile. Rutile phase was formed due to the high temperature on the surface during PEO process. Only sample that was modified with ZnO in 10min long spry pyrolysis process showed also maxima of zincite. No other phases were obtained. Thermal treatment at 500°C allowed to divest from sulphur particles that precipitates in small amount during the PEO process.

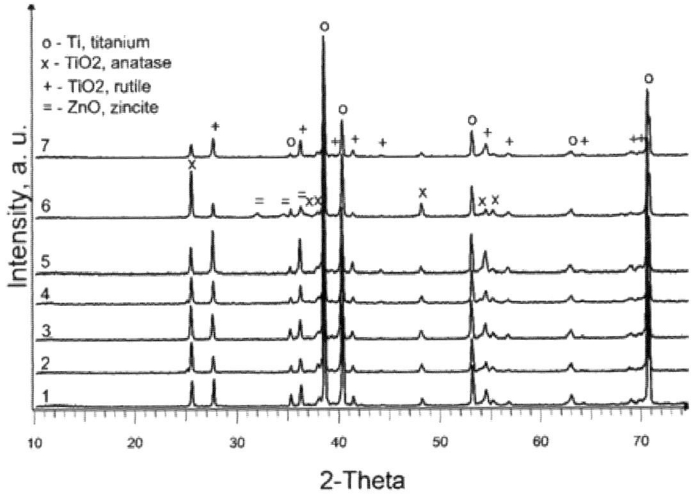

Figure 1. XRD pattern of ZnO modified porous TiO_2 coatings depending on spray pyrolysis process duration: 1) 0.5 min; 2) 1 min; 3) 3min; 4) 4 min; 5) 5min; 6) 10min; 7) pure TiO_2 porous coating.

SEM micrographs (Figure 2.) confirmed that during the PEO process porous TiO_2 coatings where obtained on surface of the titanium foil. It was noticed that micropores and also nanopores are processed in the TiO_2 layer. Some individual ZnO nanocrystals were observed on the surface of porous TiO_2 coating.

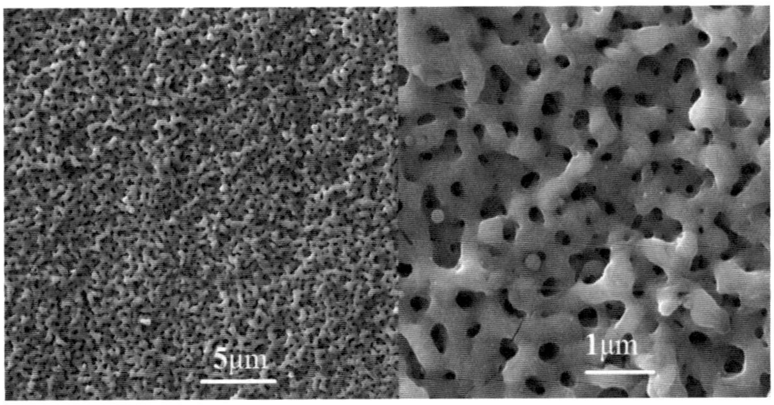

Figure 2. SEM micrographs at different magnifications of 0.14% ZnO modified TiO_2 porous coatings on titanium.

The results of photocatalytic degradation of MB are shown in Figure 3. The highest result 82.1% of degraded MB solution in 4 hours was shown by sample prepared in 0.5 min long spray pyrolysis process. The lowest photocatalytic activity shows sample prepared in 4 min long experiment and it reaches only 63.2% MB degradation rate in 4 hours. Comparing photocatalysis results with the results of XRF (Table 1.) analysis it can be notice that there is no linear connection between the amount of ZnO and the result of MB degradation. This means that the most active ZnO concentration should be investigated in experimental way. The results of XRF analysis shows that by increasing the duration of spray pyrolysis process the amount of ZnO in photocatalysts increases.

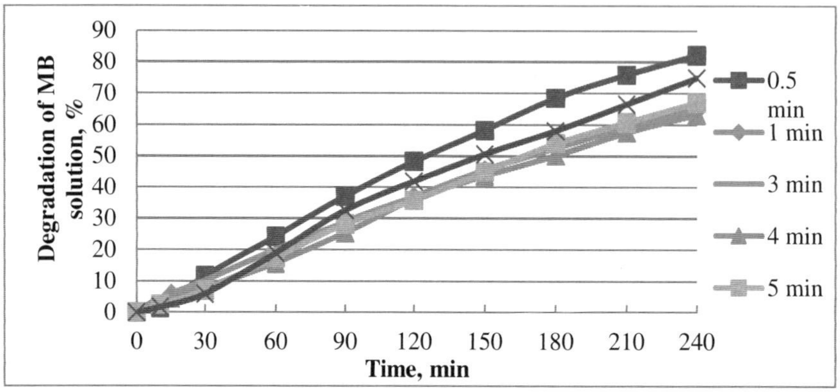

Figure 3. Photocatalytic activity of ZnO modified TiO_2 porous coatings under visible irradiation.

To estimate the kinetic rate of the MB degradation reaction, an equation form of power law is used [6]:

$$-r_{dye} = kC_{cat}mC\ n \quad (1)$$

Where k is removal rate constant, C_{dye} is dye concentration, n is reaction order respect to the dye concentration, C_{cat} is photocatalyst concentration and m is reaction order respect to the photocatalyst concentration.
As the catalyst concentration is constant, the rate equation can be stated with observed rate constant (k_{obs}) as follows:

$$-r_{dye} = -dC_{dye}/dt = k_{obs}C_{dye}n \quad (2)$$

$$k_{obs} = k.C_{cat}m \quad (3)$$

The pseudo-first order kinetic model (n=1) provide equation 3 to following form:

$$-\ln C_{dye}/C_{dye0} = k_{obs}t \quad (4)$$

$$k_{obs}= (-\ln C_{dye}/C_{dye0})/t \quad (5)$$

Results of calculated average pseudo first-rate reaction kinetic constant are shown in Table 1 and are in range of 0.581 $s^{-1}\cdot10^{-4}$ to 0.871 $s^{-1}\cdot10^{-4}$. Calculated average pseudo first-rate reaction kinetic constant values increase in the same order as photocatalytic activity results.

Table 1. Properies of obtained ZnO modified TiO_2 porous coatings depending on spray pyrolysis process duration

Duration of spry pyrolysis process, min	ZnO content, Wt%	Average pseudo first-rate reaction kinetic constant k_1, $s^{-1}\cdot10^{-4}$
0.5	0.14	0.871
1	0.18	0.635
3	0.25	0.674
4	0.36	0.581
5	0.56	0.609
10	1.26	0.677

4. Conclusions

Porous TiO_2 coatings were successfully fabricated through a simple and fast plasma electrolytic oxidation process. Obtained porous TiO_2 nano and micro porous coatings were modified with ZnO nanoparticles by using spray pyrolysis method. All samples showed strong maxima of titanium, anatase and rutile. Rutile phase was formed due to the high temperature on the surface during PEO process. Only sample that was modified with ZnO in 10min long spry pyrolysis process showed also maxima of zincite. No other phases were obtained. Modification with ZnO allows to increase photocatalytic activity of TiO_2 coatings up to 62-82% under visible light irradiation. The highest photocatalytic activity was shown by sample containing 0.14wt% ZnO that was prepared in 0.5 minutes long spray pyrolysis process. Calculated pseudo-first order kinetic constants are in range of $0.581 \cdot 10^{-4}\ s^{-1}$ to $0.871 \cdot 10^{-4}\ s^{-1}$. Prepared photocatalysts are perspective for hydrogen generation in water photocatalytic electrolysis process under visible light irradiation and also for degradation of harmful organic compounds in waste water.

Acknowledgements
This work has been supported by ERAF within the project 1.1.1.1/16/A/079. "Development of active under sunlight immobilized TiO_2-ZnO based photocatalysts".

References

[1] Banerjee S, Dionysiou D D and Pillai S C 2015 *Appl. Catal. B Environ.* **176–177** 396
[2] Fujishima A, Rao T N and Tryk D A 2000 *J. Photochem. Photobiol. C Photochem. Rev.* **1** 1
[3] Chamanzadeh Z, Noormohammadi M and Zahedifar M 2017 *Mater. Sci. Semicond. Process.* **61** 107
[4] Cheng Y, Peng Z, Wu X, Cao J, Skeldon P and Thompson G E 2015 *Electrochim. Acta* **165** 301
[5] Perednis D and Gauckler L J 2005 *J. Electroceramics* **14** 103
[6] Salehi M, Hashemipour H and Mirzaee M 2012 *Am. J. Environ. Eng.* **2** 1

IOP Conf. Series: Materials Science and Engineering **503** (2019) 012009 doi:10.1088/1757-899X/503/1/012009

Electro-catalytic and photo-catalytic reformation of CO_2– reactions and efficiencies processes (Review)

A Knoks, P Lesnicenoks, J Kleperis, L Grinberga, J Hodakovska, J Klavins, G Cikvaidze and I Lukosevics

Institute of solid state physics, University of Latvia, Kengaraga street 8, Riga, LV – 1063, Latvia

E-mail: ainars.knoks@cfi.lu.lv

Abstract. Energy harvesting with lowest environmental impact is one of key elements for cleaner future. Photocatalytic as well as electrocatalytic CO_2 reformation processes are considered as prominent methods. Thus, extensive research of CO_2 reformation is being done to find the right materials that holds crucial qualities. For photocatalysis that includes pronounced separation of light-generated opposite sign charge carriers, sensitivity to visible light, high quantum yield. In electrocatalysis high CO_2 adsorption, chemical stability, multielectron reaction catalysts are necessary. Additionally, materials participating in the reaction process must be provided with charge carriers at proper reduction and oxidation potentials. To meet the set goal of lowering environmental impact and lower CO_2 amounts exhausted into the atmosphere by human activities, it is necessary to find right technology for capturing, storing, and reusing carbon dioxide. Various technologies and materials in different levels of readiness are available and under development, such as CuO loaded TiO_2 nanotubes for photocatalytic reformation or electrocatalytic reduction on copper. Not only the proof of concept is necessary but estimation and more importantly determination of the efficiency of both electro and photo catalytic reformation of CO_2. In this work review of reactions and efficiency of both processes based on existing established technological methods is done.

1. Introduction

It is well known that proactive steps towards pollution reduction and shift from oil based energy are being pursued not only generally by legislation and agreements [1] but also by long term development strategy and research, such as EU climate-neutral plan by 2050. Not only energy harvesting with lowest environmental impact is on of key elements for cleaner future, but also use of accessible resources more efficiently. Electrocatalytic and photocatalytic CO_2 reformation are considered as prominent methods for lowering atmospheric pollution and providing possible shift to more sustainable energy production. CO_2 can be reformed into various hydrocarbons that can be directly used as fuel or substitute conventional chemical production. Extensive research on CO_2 reformation is being done to find the right materials for successful catalytic CO_2 reduction. It is known that straightforward CO_2 reduction requires substantial amount of energy and special conditions such as high pressure and temperature [2,3]. It is possible to increase efficiency of the process using catalytic reformation under lower temperatures and increase production yields. Materials participating in the reaction must possess adequate CO_2 adsorption properties, provide with charge carriers with proper reduction and oxidation potentials for successful reformation of CO_2.

Various technologies and catalysts in different technology readiness levels are available or are under development, such as CuO loaded TiO_2 nanotubes for photocatalytic reformation [4] or electrocatalytic reduction on various forms of copper [5,6]. Estimation and more importantly determination of the efficiency of both electrocatalytic and photocatalytic reformation of CO_2 is necessary.

In this work review of reactions and efficiency of both processes based on existing technological methods was done.

2. Electrocatalytic process

CO_2 electrocatalytic reduction has been under investigation for decades, but many uncertainties still remain, such as the reformation pathway from CO_2 to higher hydrocarbons, electrolyte influence, etc. The question has been tackled by deWulf and Bard looking into copper electrode using electrochemical methods [7], Firet et al. investigation of silver thin films through spectroscopic methods [8], Shen et al. investigating cobalt protoporphyrin immobilized on a pyrolytic graphite [9] or Kuhl et al. investigating various metal surfaces [10]. It has come to classifications of CO_2 reduction products into two categories: 1) gaseous; 2) liquid products, requiring different physical separation and characterization/identification methodologies. On the other hand, the reformation reactions also can be classified by the required number of electrons necessary for a product. Then the reactions can be written as

$$\mathrm{CO_2(g) + 2H^+ + 2e^- = HCOOH(l)} \quad E = -250\ \mathrm{mV} \quad (1)$$
$$\mathrm{CO_2(g) + H_2O(l) + 2e^- = CO(g) + H_2O(l)} \quad E = -1078\ \mathrm{mV} \quad (2)$$
$$\mathrm{CO_2(g) + 2H^+ + 2e^- = CO(g) + H_2O(l)} \quad E = -110\ \mathrm{mV} \quad (3)$$
$$\mathrm{CO_2(g) + 2H_2O(l) + 2e^- = CO(g) + H_2O(l)} \quad E = -930\ \mathrm{mV} \quad (4)$$
$$\mathrm{CO_2(g) + 4H^+ + 4e^- = CH_2O(l) + H_2O(l)} \quad E = -70\ \mathrm{mV} \quad (5)$$
$$\mathrm{CO_2(g) + 3H_2O(l) + 4e^- = CH_2O(l) + 4OH^-} \quad E = -900\ \mathrm{mV} \quad (6)$$
$$\mathrm{CO_2(g) + 6H^+ + 6e^- = CH_3OH(l) + H_2O(l)} \quad E = +20\ \mathrm{mV} \quad (7)$$
$$\mathrm{CO_2(g) + 5H_2O(l) + 6e^- = CH_3OH(l) + 6OH^-} \quad E = -810\ \mathrm{mV} \quad (8)$$
$$\mathrm{CO_2(g) + 8H^+ + 8e^- = CH_4(g) + 2H_2O(l)} \quad E = +170\ \mathrm{mV} \quad (9)$$
$$\mathrm{CO_2(g) + 6H_2O(l) + 8e^- = CH_4(g) + 8\ OH^-} \quad E = -660\ \mathrm{mV} \quad (10)$$
$$\mathrm{2CO_2(g) + 2H^+ + 2e^- = H_2C_2O_4(aq)} \quad E = -500\ \mathrm{mV} \quad (11)$$
$$\mathrm{2CO_2(g) + 2e^- = C_2O_4^{2-}(aq)} \quad E = -590\ \mathrm{mV} \quad (12)$$
$$\mathrm{2CO_2(g) + 12H^+ + 12e^- = CH_2CH_2(g) + 4H_2O(l)} \quad E = +60\ \mathrm{mV} \quad (13)$$
$$\mathrm{2CO_2(g) + 12H^+ + 12e^- = CH_3CH_2}OH\mathrm{(l) + 3H_2O(l)} \quad E = +80\ \mathrm{mV} \quad (14)$$
$$\mathrm{2CO_2(g) + 12H^+ + 12e^- = C_2}H_4(g)\mathrm{ + 4H_2O(l)} \quad E = +80\ \mathrm{mV} \quad (15)$$

here electrode potential E vs standard hydrogen electrode (SHE) in normal conditions [11–14].

We clearly see that hydrocarbons such as ethanol and ethylene reactions require multi-electron transfer. In addition, overpotential plays an important role in the propagation of a reaction, especially involving higher number of electrons. Other parameters, such as electrolyte composition and concentration, pH, temperature, catalyst crystalline facets, dopants, must be considered. For example, majority products on Ag is CO, Cd produces $HCOO^-$, on the other hand, various forms of Cu can produce wide variety of products [11,12,15–17]. Let's assume interest in hydrocarbons, such as C_2H_4, then appropriate catalyst will be selected by majority product, faradaic efficiency (FE), overpotential and catalyst synthesis process.

In early work of Hori et al [18] the reformation of CO to CH_2 and C_2H_4 was discussed. Process was carried out at copper electrode in aqueous electrolyte. It was reported that CO electrocatalytic reduction didn't proceed as the cathodic currents were on scale of 10^{-5} Acm^{-2} [18]. Nowadays the current densities reported are with magnitude of 10^{-2} [19] and up to 10^{1} [13].

In later work of Hori et al. it was pointed out that CO_2 must initially be reduced to CO, which is adsorbed on the catalyst. Only then further reformation to C2 and C3 products are possible. Investigation of pH showed that CH_4 production is promoted at low pH and C_2H_4 at high pH, due to increase in electrolyte concentration of $KHCO_3$. Dependency was explained as ^-OH neutralization by nearby HCO_3^-

. It is noteworthy that current densities for C_2H_4 and C_2H_5OH were reported independent of pH level [20].

Electrolyte concentration influence on FE and production of hydrocarbons was investigated by Zhong et al. where they suggested that higher electrolyte concentrations provide with lower overpotentials at the same time increasing H_2 production FE [21]. At the same time it was reported that C_2H_4 production is more favourable under low overpotentials [5]. Thus, indicating that if we are interested in C_2H_4 production, a catalyst with low overpotential must be chosen and an electrolyte with high pH level can be used, due to reported current density independence of pH. But high concentration and pH promotes H_2 production, that sometimes is considered the rate determining step.

It was proposed as H_2^+ (ads) or $H_{ads} + H^+$ in comparison with previously assumed $H_2^+(ads) + e^- \rightarrow H_2$, where H_2^+ (ads) is transition state. If carbon adsorption on electrode followed by OH electronation and not H_2^+ (ads) or $H_{ads} + H^+$ is the rate determining step for ethylene, a formation diagram was proposed (See Figure 1) [20].

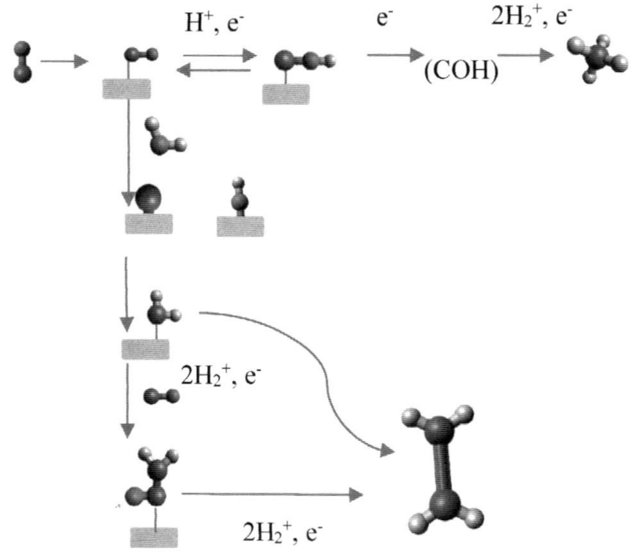

Figure 1. Reaction pathway of CO_2 reduction with intermediate adsorbed CO, COH, C and OH. Adopted from [20], molecules were drawn by using [24].

In case of pathway depicted in Figure 1, the rate determining step can be CO adsorption and following reformation into CH_4 in addition to various intermediates. It is noteworthy that other authors also came to similar conclusions and results for Cu nanoparticles [22,23].

Kortlever et al. suggested that CO_2 reduction can be viewed as multiple proton-electron reactions with products C1, C2+ and water, summarizing all probable product equations as equation (16)

$$kCO_2 + n(H^+ + e^-) \leftrightarrows P + mH_2O \quad (16)$$

where P is the C1 and C2+ products [12].

Reaction involving more than 2 electrons (and protons) will involve intermediate steps, thus it will cause catalytic irreversibility. The role of intermediates in the CO_2 reduction was discussed. For example, ethylene formation starts with *CO and *CH_2 intermediates, which follows abovementioned pathway [12]. But during investigation three unexplained observations were made. First, formation of methane from CO shows different pH dependency compared to ethylene. Second, often ethylene formation happened at more positive (less negative) potential without methane formation. Third, there is a structure sensitive and pH dependent pathways on Cu. Cu(111) reduction of CO to methane and ethylene is observed simultaneously, but on Cu(100) different pathway is observed, where only ethylene is formed, especially at high pH values [12]. On the electrode surface due to CO_2 reduction there is a pH gradient that influences product selectivity. The pH value and ion presence can stabilize reaction intermediates, thus promoting formation of one or the other product [25]. Due to proton consumption during the reduction process and hydrogen evolution reaction (HER) the local pH at the electrode deviates from bulk, thus CO_2 can react with OH^- and form bicarbonate or carbonate [12]. On the other hand, Resasco et al. did an extensive investigation on production of CO, $HCOO^-$, C_2H_4 and CH_3CH_2OH on Cu electrode, and concluded that the concentration of electrolyte anions has very little influence on production. Attributing it to lack of involvement of hydrogen atoms. At the same time providing with evidence for hydrogen and CH_4 production high sensitivity to anion concentration [26].

Malik et al [27] described electrochemical reduction of CO_2 process on various Cu morphologies, with FE of 26% at -2.75V. It was reported that grown dendrites on the surface capture some of gaseous products lowering overall efficiency. In addition, the use of Cu nanoparticles increased CO production, with selectivity dependence on size [27,28] attributing it to low-coordination site activity. Similar results concerning increase in CO production was found by Reske et al. decrease in particle size promotes H_2 and CO production, at the same time providing with higher current density in comparison with bulk Cu [28].

Authors investigation CO_2 reformation on metals as Fe, Au, Zn, Ni, Pt and Ag propose similar reformation pathway [29,30] as seen in Figure 1. Leading to conclusion that first step in CO_2 reformation is adsorption of CO_2, then CO is formed and desorbed or transformed into *CHO or *CHO species, that later forms into hydrocarbons as methane and methanol. It is noteworthy that metals as Pt, Pd, and Ni were reported to bind CO too strongly, which leads to electrode poisoning by *CO [25,31]. As previously mentioned crystallographic facets show selectivity for production: Cu (111) for CH_4 and Cu (100) for C_2H_4, suggesting that CO adsorption and lower surface energy is the reason of selectivity [27,32,33]. An overview of FE of different materials in various formations is presented in Table 1.

Table 1. FE and major product summary NP – nanoparticle, NW – nanowire, D – dendrite, NR – nanorods, NC – nanocubes, NF – nanoflowers,

Ref	Catalyst	electrolyte	pH	U, V vs RHE	FE	Product	Ref nr
Chen et al	Cu0 - Cu_2O	0.1M $KHCO_3$	6.8	-1	32.1	C_2H_4	[34]
Chen et al	Cu0 - Cu_2O	0.1M $KHCO_3$	6.8	-1	16.4	C_2H_5OH	[34]
Chen et al	Cu_2O-PdClx	0.1M $KHCO_3$	6.8	-1	30.1	C_2H_6	[34]
Chen et al	Graphite-$PdCl_2$	0.1M $KHCO_3$	6.8	-1	4.7	$HCOO^-$	[34]
Chen et al	Graphite-$PdCl_2$	0.1M $KHCO_3$	6.8	-1	8.1	CO	[34]
Chen et al	Graphite-$PdCl_2$	0.1M $KHCO_3$	6.8	-1	0.2	CH_4	[34]
Hori et al	Cu	KCl	5.9	-1.44	47.8	C_2H_4	[5]
Hori et al	Cu	$KHCO_3$	6.8	-1.41	30.1	C2H4	[5]
Tang et al	Cu NP				30	CO	[35]
Tang et al	Cu NP				35	C_2H_4	[35]
Reske et al	Cu NP	0.1 M $KHCO_3$	6.8	-1.1 V	20/70	CO/H_2	[28]
Raciti et al	Cu NW			-0.6 V	30.7	HCOOH	[36]
Raciti et al	Cu NW	0.1M $KHCO_3$		-0.4 V	60	CO	[36]
Chen et al	Cu mesocrystal	0.1M $KHCO_3$		-0.99 V	81	C_2H_4	[37]
Roberts et al	Cu NC	0.1M $KHCO_3$		-0.6 V		C_2H_6	[38]
Sen et al	Cu foam	0.5M $KHCO_3$		-0.5 V	26	HCOOH	[39]
Sen et al	Cu smooth	0.5M $KHCO_3$		-0.5 V	3	HCOOH	[39]
Sen et al	Cu foam	0.5M $KHCO_3$		-0.9 V	37	HCOOH	[39]
Kas et al		0.1 M $KHCO_3$	6.8	-1.1 V	40	C_2H_6	[40]
Malik et al	Cu NR	0.5 M $KHCO_3$		-2.75 V	24.5	CO	[27]

Malik et al	Cu D	0.5 M $KHCO_3$		-3 V	7	C_2H_6	[27]
Malik et al	Cu D	0.5 M $KHCO_3$		-2.75 V	7	C_2H_4	[27]
Malik et al	Cu octahedral	0.5 M $KHCO_3$		-3,0 V	5	CO	[27]
Malik et al	Cu D	0.5 M $KHCO_3$		-2,5 V	2,8	CH_4	[27]
Lui et al	Sn/Sn_{O2}	0.5 M $KHCO_3$	7.2	-1.4 V	70	$HCOO^-$	[41]
Mistry et al	Au/Cu NP	0.1 M $KHCO_3$	6.8	-1.2 V	20	CO/CH_4	[42]
Wang et al	Cu_2Cd/Cd/Cu	0.1 M $KHCO_3$		-1.0 V	84	CO	[43]
Bernal et al	CuCo NP	0.1 M $KHCO_3$		-1.1 V	11	HCOOH	[44]
Bernal et al	CuCo NP	0.1 M $KHCO_3$		-1.1 V	7.5	CO	[44]
Yang et al	Co NF	0.1 M $KHCO_3$		-0.456 V	63.4	Formate	[45]
Ge et al	Sn_{O2}			-0.8 V	80	CO/Formate	[46]
Kaneco et al	Cu NP with Pb	NaOH in CH_3OH	14	-2.3 V	80	HCOOH	[47]
Kaneco et al	Cu NP with Zn	NaOH in CH_3OH	14	-2.1 V	12	CH_4	[47]
Cook et al	Cu on GDE	1 M KOH	14	- 2.8 V	69	C_2H_4	[48]

Particle size also influence not only selectivity but also FE. CuCo NP showed decrease in HER, with increase of Cu content NP. On the other hand increased Cu content lead to increase in production of CO and HCOOH [44]. Flower like structures were proven to improve production of formate up to 63% FE. From Table 1 we can see that Cu is promising material for CO_2 reformation. Similar results were presented by Zhang et al. not only for Cu, but also for Zn and Sn based catalysts [49].

3. Photocatalytic CO_2 reduction

One of the problems in photocatalytic CO_2 reduction is that hydrogen donors, such as H_2O, CH_4, and alcohols, act as hole scavengers, thus it is necessary to separate reduction products from bulk electrolyte. This correlates with hydrogen production problems. Assuming that the CO_2 reduction reaction can proceed without limits set by charge carrier potentials (valence band (VB) and conduction band (CB) band position) [50].

Electrons are excited into the CB, with that, a hole is created in VB, where these charge carriers could be trapped in shallow or deep sites. In TiO_2 it could be $Ti^{III}OH$ and $Ti^{IV}OH^{\bullet}$ shallow sites. Ti^{III} also could trap electrons in bulk material. If no site is found, charge carriers recombine and create excess heat on the material surface. Hydrocarbons can be produced only if hydride reductants are oxidized simultaneously. Thus, there is a need of oxidizing agents and oxidizing mechanism [50,51].

Direct oxidizing species are holes, but there are additional molecules with similar oxidizing properties: hydroxyl radicals ($^{\bullet}OH_{free}$ and $^{\bullet}OH_{ads}$ [52]), superoxide ($^{\bullet}O_2^-$ [53]), singlet oxygen ($^{-}O_2$ [54]), also some authors report hydrogen peroxide and oxygen [55]. Karamian et al summarizes that $^{\bullet}OH$ radicals are mostly product of surface hydroxyl and adsorbed water oxidation process. In this process peroxide can be produced and oxidized as described by equations (17) to (21).

Superoxides are supposed to be a weak oxidizing species as a product of one CB electron reduction of molecular oxygen or oxidizing hydrogen peroxide by VB hole (equation 22 to 24), whereas single oxygen is a strong oxidant and can be a product of $\bullet O_2^-$ and a trapped hole, though it has a short lifetime 2 μs on the surface, especially compared with hydroxyl radical with lifetime 5 times longer [50,54,55].

$$H_2O_2 + h\nu \rightarrow \bullet OH_{free/ads} + \bullet OH_{free/ads} \quad (17)$$
$$H_2O_2 + e_{CB}^- \rightarrow \bullet OH_{ads} + OH^- \quad (18)$$
$$H_2O_2 + \bullet O_2^- \rightarrow \bullet OH_{free/ads} + OH^- + O_2 \quad (19)$$
$$H_2O_2 + h_{VB}^+ + 2OH^- \rightarrow \bullet O_2^- + 2H_2O \quad (20)$$
$$H_2O_2 + \bullet OH + OH^- \rightarrow \bullet O_2^- + 2H_2O \quad (21)$$

$$OH^- + h_{VB}^+ \rightarrow \ ^{\bullet}OH_{ads} \quad (22)$$
$$O_2 + e_{CB}^- \rightarrow \ ^{\bullet}O_2^- \quad (23)$$
$$O_2 + 2e_{CB}^- + 2H^+ \rightarrow H_2O_2 \quad (24)$$

It must be remembered that CO_2 is very stable molecule, difference between lowest unoccupied molecular orbital (LUMO) and highest occupied molecular orbital states is 13.7 eV. Adsorption of CO_2 onto catalyst is the best way to lower its energy due to the bend in LUMO level. One electron transfer initiates the adsorption and following chemical reactions. Electron or proton transfer provides with C-O bond breaking and C-H bond creation. It is believed that capturing electron from CB of photocatalyst is the first step in $\bullet CO_2^-$ anion radical as described in the following equation (25). Additional reductant is necessary, it is supposed that carbon-free reductants such as H_2 or water allow C1 products and carbon containing reductants such as CH_4 can promote C2, C3 products [50].

$$CO_2 + e_{CB}^- \rightarrow \cdot CO_2^- \quad (25)$$

P-type photocathodes and earth-abundant n-type anodes such as WO_3, Fe_2O_3, TiO_2 can be used for photocatalytic CO_2 reduction. By utilizing metallic and composite catalysts as cathodes one can gain high FE and products as CH_4 and C_2H_4. In a photocatalytic cell using WO_3 and Cu or Sn/SnO_x electrodes CO_2 reduction potential of -3.0 V and +0.2V *vs* Reversible hydrogen electrode (RHE) has been reported. Thus the photocatalytic CO_2 reduction is described by equations (1)-(15) with lower bias potential due to photocatalytic charge carrier generation [56]. FE and necessary bias potentials for photocatalytic CO_2 reduction are provided in Table 2.

Table 2 Summary of FE and bias potentials in photocatalytic CO_2 reduction.

Ref	Catalyst	electrolyte	pH	Potential V vs RHE	FE	Product	Ref nr	λ, nm
Magesh et al	WO_3-Cu	$KHCO_3$	7.5	+0.87	42.3	CH_4	[57]	>420
'-'	WO_3-Cu	$KHCO_3$	7.5	+0.65	64.6	CH_4	[57]	>420
'-'	WO_3-Cu	$KHCO_3$	7.5	+0.75	67	CH_4	[57]	>420
'-'	WO_3-Cu	$KHCO_3$	7.5	+0.87	4	C_2H_4	[57]	>420
'-'	WO_3-Cu	$KHCO_3$	7.5	+0.65	4.5	C_2H_4	[57]	>420
'-'	WO_3-Cu	$KHCO_3$	7.5	+0.75	2.7	C_2H_4	[57]	>420
'-'	WO_3-Sn/SnO_x	KCl	7.5	+0.6	27.5	HCOOH	[57]	>420
'-'	WO_3-Sn/SnO_x	KCl	7.5	+0.7	27.5	HCOOH	[57]	>420
'-'	WO_3-Sn/SnO_x	KCl	7.5	+0.8	26.8	HCOOH	[57]	>420
Hinogami et al	p-Si	0.1 M $KHCO_3$	6.8	-0.62	12	CO	[58]	>420
'-'	Cu/p-Si	0.1 M $KHCO_3$	6.8	-0.52	25	CH_4	[58]	>420
'-'	p-Si	0.1 M $KHCO_3$	6.8	-1.026	73.5	H_2	[58]	>420
'-'	Ag	0.1 M $KHCO_3$	6.8	-1.49	75.9	CO	[58]	>420
'-'	part-Ag/p-Si	0.1 M $KHCO_3$	6.8	-0.806	50.9	CO	[58]	>420
'-'	Au	0.1 M $KHCO_3$	6.8	-0.966	82.2	CO	[58]	>420
'-'	part Au/p-Si	0.1 M $KHCO_3$	6.8	-0.496	62.2	CO	[58]	>420
Yoneyama et al	p-CdTe	0.1 M Na_2CO_3	n.a.	-0.55	48.4	HCOOH	[59]	>450
'-'	p-CdTe	0.1 M TEAP	n.a.	-0.55	65.6	CO	[59]	>450
'-'	p-InP	0.1 M TEAP	n.a.	-0.55	17.5	HCOOH	[59]	>450
'-'	p-InP	0.1 M Na_2CO_3	n.a.	-0.55	36.2	HCOOH	[59]	>450
Tan et al	5GO-OTiO_2	Gaseous CO_2 with H_2O	n.a.	0	3.45	CH_4	[60]	>400

‘’-‘’	5GO-OTiO_2	Gaseous CO_2 with H_2O	n.a.	0	9.07	CO	[60]	>400
‘’-‘’	5GO-OTiO_2	Gaseous CO_2 with H_2O	n.a.	0	1.75	C_2H_4	[60]	>400

4. Conclusions

Catalytic CO_2 reformation is yet to be fully understood. The pathways of electrocatalytic and photocatalytic reduction are similar but requires different parameter control and thus have different production efficiency.

In electrocatalytic reduction higher hydrocarbons are produced under more negative overpotentials than -1.3 V vs NHE. Low carbonate and KCl concentrations provide higher FE for C_2H_4 formation. pH influence is not determined but results suggest that flow of electrolyte should be applied for higher FE.

Electrocatalytic CO_2 reduction in liquid electrolytes is characterized by higher production rates and photocatalytic reduction production volumes are substantially lower. Production of higher hydrocarbons is challenging in photocatalytic reduction, as the majority of products are C1 and C2. In addition, in many cases photocatalytic CO_2 reduction requires bias potential, which increases overall energy consumption. It is necessary to investigate materials that would allow CO_2 reduction without bias potential. Photocatalytic processes are not able to deliver high enough current densities for high volume production.

If the goal is C_2H_4 then nanostructured Cu should be chosen as catalyst material, higher concentration electrolyte should be used in electrocatalytic CO_2 reformation.

5. Acknowledgement

Funding from European Union`s Horizon 2020 Research and Innovation Program project under grant agreement No 768789 is greatly acknowledged.

References

[1] Anon 2016 *Communication from the commission to the European parliament, the council, the European economic and social committee and the committee of the regions: a European strategy for low-emission mobility 2016* (Brussels: European Commissio)

[2] Alper E and Yuksel Orhan O 2017 CO_2 utilization: Developments in conversion processes *Petroleum* **3** 109–26

[3] Whipple D T and Kenis P J A 2010 Prospects of CO_2 utilization via direct heterogeneous electrochemical reduction *J. Phys. Chem. Lett.* **1** 3451

[4] Razali M H and Yusoff M 2018 Highly efficient CuO incorporated TiO_2 Nanotube photocatalyst for hydrogen production from water *Mater. Lett.* 168–71

[5] Hori Y, Murata A and Takahashi R 1989 Formation of hydrocarbons in the electrochemical reduction of carbon dioxide at a copper electrode in aqueous solution *J. Chem. Soc. Faraday Trans. 1 Phys. Chem. Condens. Phases* **85** 2309

[6] Gunasekar G H, Shin J, Jung K-D, Park K and Yoon S 2018 Design Strategy toward Recyclable and Highly Efficient Heterogeneous Catalysts for the Hydrogenation of CO_2 to Formate *ACS Catal.* 4346–4353

[7] DeWulf D W, Jin T and Bard A J 1989 Electrochemical and Surface Studies of Carbon Dioxide Reduction to Methane and Ethylene at Copper Electrodes in Aqueous Solutions *J. Electrochem. Soc.* **136** 1686

[8] Firet N J and Smith W A 2017 Probing the Reaction Mechanism of CO_2 Electroreduction over Ag Films via Operando Infrared Spectroscopy *ACS Catal.* **7** 606–12

[9] Shen J, Kortlever R, Kas R, Birdja Y Y, Diaz-Morales O, Kwon Y, Ledezma-Yanez I, Schouten K J P, Mul G and Koper M T M 2015 Electrocatalytic reduction of carbon dioxide to carbon

monoxide and methane at an immobilized cobalt protoporphyrin *Nat. Commun.* **6** 1–8
[10] Kuhl K P, Hatsukade T, Cave E R, Abram D N, Kibsgaard J and Jaramillo T F 2014 Electrocatalytic Conversion of Carbon Dioxide to Methane and Methanol on Transition Metal Surfaces *J. Am. Chem. Soc.* **136** 14107–13
[11] Qiao J, Liu Y, Hong F and Zhang J 2014 A review of catalysts for the electroreduction of carbon dioxide to produce low-carbon fuels *Chem. Soc. Rev.* **43** 631–75
[12] Kortlever R, Shen J, Schouten K J P, Calle-Vallejo F and Koper M T M 2015 Catalysts and Reaction Pathways for the Electrochemical Reduction of Carbon Dioxide *J. Phys. Chem. Lett.* **6** 4073–82
[13] Francke R, Schille B and Roemelt M 2018 Homogeneously Catalyzed Electroreduction of Carbon Dioxide - Methods, Mechanisms, and Catalysts *Chem. Rev.* **118** 4631–70
[14] Townshend A 1987 Standard potentials in aqueous solutions *Anal. Chim. Acta* **198** 333–4
[15] Hori Y, Kikuchi K and Suzuki S 1985 Production of CO and CH_4 in electochemical reduction of CO_2 at metal electrodes in aqueous hydrogencarbonate solution *Chem. Lett.* **14** 1695–8
[16] Jitaru M, Lowy D A, Toma M, Toma B C and Oniciu L 1997 Electrochemical reduction of carbon dioxide on flat metallic cathodes *J. Appl. Electrochem.* **27** 875–89
[17] Gattrell M, Gupta N and Co A 2006 A review of the aqueous electrochemical reduction of CO_2 to hydrocarbons at copper *J. Electroanal. Chem.* **594** 1–19
[18] Hori Y, Murata A, Takahashi R and Suzuki S 1987 Electroreduction of carbon monoxide to methane and ethylene at a copper electrode in aqueous solutions at ambient temperature and pressure *J. Am. Chem. Soc.* **109** 5022–3
[19] Zhang Q, Xu W, Xu J, Liu Y and Zhang J 2018 High performing and cost-effective metal/metal oxide/metal alloy catalysts/electrodes for low temperature CO_2 electroreduction *Catal. Today* 0–1
[20] Hori Y, Takahashi R, Yoshinami Y and Murata A 1997 Electrochemical Reduction of CO at a Copper Electrode *J. Phys. Chem. B* **101** 7075–81
[21] Zhong H, Fujii K and Nakano Y 2017 Effect of $KHCO_3$ Concentration on Electrochemical Reduction of CO_2 on Copper Electrode *J. Electrochem. Soc.* **164** F923–7
[22] Ma S, Sadakiyo M, Luo R, Heima M, Yamauchi M and Kenis P J A 2016 One-step electrosynthesis of ethylene and ethanol from CO_2 in an alkaline electrolyzer *J. Power Sources* **301** 219–28
[23] Kuhl K P, Cave E R, Abram D N and Jaramillo T F 2012 New insights into the electrochemical reduction of carbon dioxide on metallic copper surfaces *Energy Environ. Sci.* **5** 7050
[24] Hanwell M D, Curtis D E, Lonie D C, Vandermeerschd T, Zurek E and Hutchison G R 2012 Avogadro: An advanced semantic chemical editor, visualization, and analysis platform *J. Cheminform.* **4** 17
[25] Hori Y 2008 Electrochemical CO_2 Reduction on Metal Electrodes *Modern Aspects of Electrochemistry* (New York, NY: Springer New York) pp 89–189
[26] Resasco J, Lum Y, Clark E, Zeledon J Z and Bell A T 2018 Effects of Anion Identity and Concentration on Electrochemical Reduction of CO_2 *ChemElectroChem* **5** 1064–72
[27] Malik K, Bajaj N K and Verma A 2018 Effect of catalyst layer on electrochemical reduction of carbon dioxide using different morphologies of copper *J. CO_2 Util.* **27** 355–65
[28] Reske R, Mistry H, Behafarid F, Roldan Cuenya B and Strasser P 2014 Particle size effects in the catalytic electroreduction of CO_2 on Cu nanoparticles *J. Am. Chem. Soc.* **136** 6978–86
[29] Schouten K J P, Kwon Y, van der Ham C J M, Qin Z and Koper M T M 2011 A new mechanism for the selectivity to C1 and C2 species in the electrochemical reduction of carbon dioxide on copper electrodes *Chem. Sci.* **2** 1902
[30] Hatsukade T, Kuhl K P, Cave E R, Abram D N and Jaramillo T F 2014 Insights into the electrocatalytic reduction of CO_2 on metallic silver surfaces *Phys. Chem. Chem. Phys.* **16** 13814–9
[31] Hori Y, Wakebe H, Tsukamoto T and Koga O 1994 Electrocatalytic process of CO selectivity in

electrochemical reduction of CO_2 at metal electrodes in aqueous media *Electrochim. Acta* **39** 1833–9

[32] Hori Y, Takahashi I, Koga O and Hoshi N 2003 Electrochemical reduction of carbon dioxide at various series of copper single crystal electrodes *J. Mol. Catal. A Chem.* **199** 39–47

[33] Huang Y, Handoko A D, Hirunsit P and Yeo B S 2017 Electrochemical Reduction of CO_2 Using Copper Single-Crystal Surfaces: Effects of CO* Coverage on the Selective Formation of Ethylene *ACS Catal.* **7** 1749–56

[34] Chen C S, Wan J H and Yeo B S 2015 Electrochemical Reduction of Carbon Dioxide to Ethane Using Nanostructured Cu2O-Derived Copper Catalyst and Palladium(II) Chloride *J. Phys. Chem. C* **119** 26875–82

[35] Tang W, Peterson A A, Varela A S, Jovanov Z P, Bech L, Durand W J, Dahl S, Nørskov J K and Chorkendorff I 2012 The importance of surface morphology in controlling the selectivity of polycrystalline copper for CO_2 electroreduction *Phys. Chem. Chem. Phys.* **14** 76–81

[36] Raciti D, Livi K J and Wang C 2015 Highly Dense Cu Nanowires for Low-Overpotential CO_2 Reduction *Nano Lett.* **15** 6829–35

[37] Chen C S, Handoko A D, Wan J H, Ma L, Ren D and Yeo B S 2015 Stable and selective electrochemical reduction of carbon dioxide to ethylene on copper mesocrystals *Catal. Sci. Technol.* **5** 161–8

[38] Roberts F S, Kuhl K P and Nilsson A 2015 High selectivity for ethylene from carbon dioxide reduction over copper nanocube electrocatalysts *Angew. Chemie - Int. Ed.* **54** 5179–82

[39] Sen S, Liu D and Palmore G T R 2014 Electrochemical reduction of CO_2 at copper nanofoams *ACS Catal.* **4** 3091–5

[40] Kas R, Kortlever R, Milbrat A, Koper M T M, Mul G and Baltrusaitis J 2014 Electrochemical CO_2 reduction on Cu_2O-derived copper nanoparticles: Controlling the catalytic selectivity of hydrocarbons *Phys. Chem. Chem. Phys.* **16** 12194–201

[41] Liu S, Pang F, Zhang Q, Guo R, Wang Z, Wang Y, Zhang W and Ou J 2018 Stable nanoporous Sn/SnO_2 composites for efficient electroreduction of CO_2 to formate over wide potential range *Appl. Mater. Today* **13** 135–43

[42] Mistry H, Reske R, Strasser P and Roldan Cuenya B 2017 Size-dependent reactivity of gold-copper bimetallic nanoparticles during CO_2 electroreduction *Catal. Today* **288** 30–6

[43] Wang C, Cao M, Jiang X, Wang M and Shen Y 2018 A catalyst based on copper-cadmium bimetal for electrochemical reduction of CO_2 to CO with high faradaic efficiency *Electrochim. Acta* **271** 544–50

[44] Bernal M, Bagger A, Scholten F, Sinev I, Bergmann A, Ahmadi M, Rossmeisl J and Cuenya B R 2018 CO_2 Electroreduction on Copper-Cobalt Nanoparticles: Size and Composition Effect *Nano Energy* **53** 27–36

[45] Yang G, Yu Z, Zhang J and Liang Z 2018 A highly efficient flower-like cobalt catalyst for electroreduction of carbon dioxide *Chinese J. Catal.* **39** 914–9

[46] Ge H, Gu Z, Han P, Shen H, Al-Enizi A M, Zhang L and Zheng G 2018 Mesoporous tin oxide for electrocatalytic CO_2 reduction *J. Colloid Interface Sci.* **531** 564–9

[47] Kaneco S, Ueno Y, Katsumata H, Suzuki T and Ohta K 2006 Electrochemical reduction of CO_2 in copper particle-suspended methanol *Chem. Eng. J.* **119** 107–12

[48] Cook R L 1990 High Rate Gas Phase CO_2 Reduction to Ethylene and Methane Using Gas Diffusion Electrodes *J. Electrochem. Soc.* **137** 607

[49] Zhang Q, Xu W, Xu J, Liu Y and Zhang J 2018 High performing and cost-effective metal/metal oxide/metal alloy catalysts/electrodes for low temperature CO_2 electroreduction *Catal. Today*

[50] Karamian E and Sharifnia S 2016 On the general mechanism of photocatalytic reduction of CO_2 *J. CO2 Util.* 194–203

[51] Li K, An X, Park K H, Khraisheh M and Tang J 2014 A critical review of CO_2 photoconversion: Catalysts and reactors *Catal. Today* **224** 3–12

[52] Xiang Q, Yu J and Wong P K 2011 Quantitative characterization of hydroxyl radicals produced

by various photocatalysts *J. Colloid Interface Sci.* **357** 163–7
[53] Hirakawa T, Yawata K and Nosaka Y 2007 Photocatalytic reactivity for O_2- and OH radical formation in anatase and rutile TiO_2 suspension as the effect of H_2O_2 addition *Appl. Catal. A Gen.* **325** 105–11
[54] Nosaka Y, Daimon T, Nosaka A Y and Murakami Y 2004 Singlet oxygen formation in photocatalytic TiO_2 aqueous suspension *Phys. Chem. Chem. Phys.* **6** 2917
[55] Fujishima A, Zhang X and Tryk D 2008 TiO_2 photocatalysis and related surface phenomena *Surf. Sci. Rep.* **63** 515–82
[56] Mikkelsen M, Jørgensen M and Krebs F C 2010 The teraton challenge. A review of fixation and transformation of carbon dioxide *Energy Environ. Sci.* **3** 43–81
[57] Magesh G, Kim E S, Kang H J, Banu M, Kim J Y, Kim J H and Lee J S 2014 A versatile photoanode-driven photoelectrochemical system for conversion of CO_2 to fuels with high faradaic efficiencies at low bias potentials *J. Mater. Chem. A*
[58] Hinogami R, Nakamura Y, Yae S and Nakato Y 1998 An Approach to Ideal Semiconductor Electrodes for Efficient Photoelectrochemical Reduction of Carbon Dioxide by Modification with Small Metal Particles *J. Phys. Chem. B* **102** 974–80
[59] Yoneyama H, Sugimura K and Kuwabata S 1988 Effects of electrolytes on the photoelectrochemical reduction of carbon dioxide at illuminated p-type cadmium telluride and p-type indium phosphide electrodes in aqueous solutions *J. Electroanal. Chem.* **249** 143–53

IOP Conf. Series: Materials Science and Engineering **503** (2019) 012010 doi:10.1088/1757-899X/503/1/012010

Carbon nanotubes vs graphene nanoplatelets for 3D-printable composites

G. Gorokhov[1], D. Bychanok[1,2], D. Meisak[1], I. Shlyk[1], A. Liubimau[3], P. Angelova[4], C. Menseidov[4], E. Ivanov[4], R. Kotsilkova[4], M. Casa[5], P. Ciambelli[5] and P. Kuzhir[1,2]

[1]Research Institute for Nuclear Problems, Belarusian State University, Minsk, Belarus
[2]Tomsk State University, Tomsk, Russia
[3]Belarusian State Technological University, Minsk, Belarus
[4]OLEM, Institute of Mechanics Bulgarian Academy of Sciences, Sofia, Bulgaria
[5]Narrando Srl, Fisciano, Italy

E-mail: glebgorokhov@yandex.ru

Abstract. Polymer-based composites with nanocarbon fillers are of great interest for the wide application range including the needs of wireless communication and the development of precise measuring means and medical devices. However, the composite properties such as excellent electromagnetic energy dissipation or tailorable conductivity are not enough to solve practical problems in engineering. To be fully applicable, the composite material must be low-cost and suitable for conventional methods of fabrication, for example 3D-printing. In current research the electromagnetic properties of PLA-based composites with graphene nanoplatelets and multiwall carbon nanotubes were investigated in microwave frequency range. The synergistic effect of two fillers was observed, the investigated materials proved to be prospective for 3D-printable composite production for electromagnetic applications such as fabrication of complex geometry microwave shields and antennas.

1. Introduction

Carbon nanofillers such as multiwall carbon nanotubes (MWCNT) or graphene nanoplatelets (GNP) of low concentrations can change the thermal, mechanical [1] and electromagnetic properties [2, 3, 4, 5] of a composite making it suitable for many practical applications [6]. Nevertheless the use of nanocomposites in industry is very limited due to the processing complexity and expensiveness.

The current research is focused on the application of conventional polymer technologies to nanocomposites development with an intent to overcome the technological limitations and bring carbon nanocomposite materials closer to industrial applications. It is obvious that besides the target properties (conductivity, mechanical strength, etc.) a new material must be compatible with conventional processing methods. For instance, additive technologies are perfect for creation of complex structures for electromagnetic applications [7, 8]. The fused deposition modeling (FDM) 3D-printing technology is the most popular among the diversity of additive technologies. The working principle of the FDM 3D-printer is simple: the polymer filament is extruded through the heated nozzle placed on moving head and deposited layer by layer forming a part. Due to its simplicity and low cost the FDM printers are widely used around the world for prototyping and creation of custom parts and devices.

The most popular materials for FDM 3D-printing are polymers: Nylon, acrylonitrile-butadiene-styrene (ABS), polylactic acid (PLA) which can be used as a dielectric matrix for a composite material. The sufficient impact of nanocarbon fillers such as MWCNT to the permittivity ε of a composite allows to reach the desirable ε values properties with minimal necessary filler content and, thus, minimize the deterioration of mechanical properties. High filler contents can cause composite fragility and lead to the growth of melted material viscosity. The latter plays an important role in FDM printing process: the problem with flow caused by high viscosity can lead to an instability in layer forming. Thus the minimization of filler content is an important problem for further application of 3D-printable composites. Besides the intrinsic properties of filler particles, the effect of filler addition depends on its distribution inside the polymer. The spatial dispersion of filler particles can be affected by the agglomeration, polymer matrix properties and technical process peculiarities [9]. One of possible ways to optimize the filler content is the use of a synergistic effect between two fillers with different properties.

2. Theory

The standard approach for relatively low concentrations of nanoscale fillers inside of a dielectric matrix the Maxwell Garnett (MG) theory:

$$\varepsilon_{eff} = \varepsilon_m + \sum_{filler} \frac{1/3\sum_{i=a,b,c} n\alpha_i/V}{1 - 1/3\sum_{i=a,b,c} \frac{N_i n\alpha_i/V}{\varepsilon_m}}, \tag{1}$$

where ε_m — is the polymer matrix permittivity, n — volume concentration of filler, represented as ellipsoid with semiaxes a, b, c; α_i and N_i — are ellipsoid polarizability and depolarization factor in direction i respectively. Such a representation of filler particles as randomly oriented ellipsoids with intrinsic polarizability is conventional for composites with nanoscale filler modeling [10, 5]. It is important to understand that for real composite the ellipsoid in MG model stands not for actual nanoparticles with their specific parameters, but for effective particle which can be an agglomerate or aggregate. Thus without exact knowledge about the inner structure of the composite equation 1 may be represented as:

$$\varepsilon_{eff} = \varepsilon_m + \sum_{filler} \varepsilon_{add\ filler}, \tag{2}$$

where $\varepsilon_{add\ filler}$ — is an addition to effective complex dielectric permittivity made by specific amount of filler. In case of two non-interacting fillers A and B the electromagnetic response will be:

$$\varepsilon_{eff} = \varepsilon_m + \varepsilon_{add\ A} + \varepsilon_{add\ B}. \tag{3}$$

In this case the values of ε_m, $\varepsilon_{add\ A}$ and $\varepsilon_{add\ B}$ can be evaluated from the experimental data obtained from the measurements of empty matrix and composites containing only fillers A and B. In order to account the possible interaction between fillers and following synergistic effect the ε_{syn} addition can be introduced as:

$$\varepsilon_{eff} = \varepsilon_m + \varepsilon_{add\ A} + \varepsilon_{add\ B} + \varepsilon_{syn}. \tag{4}$$

3. Materials & Methods

To check the synergy in composites with nanocarbon the set of PLA-based (Ingeo PLA 3D850 was used) composites with industrial MWCNT and GNP (TimesNano, China) fillers was made. As the main aim of the research is to come up with an industrially applicable technology, the conventional method of melt mixing was used. Twin screw extruder was used to distribute fillers inside the polymer and obtain two masterbatches with 6 wt. % of GNP and MWCNT concentrations respectively. Composites with lower filler contents and combinations of two fillers were obtained by masterbatches dilution with pure PLA. Samples for measurements were then hot-pressed in plane layers of ~1 mm thickness.

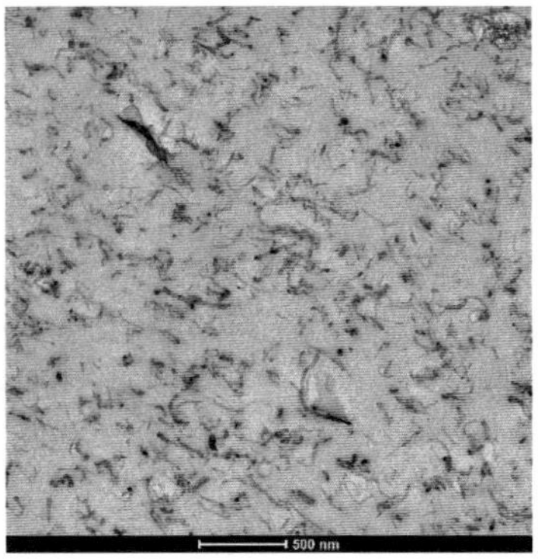

Figure 1. TEM image of composite containing 1.5 wt. % of GNP and 4.5 wt. % of MWCNT. The most of particles are well-dispersed

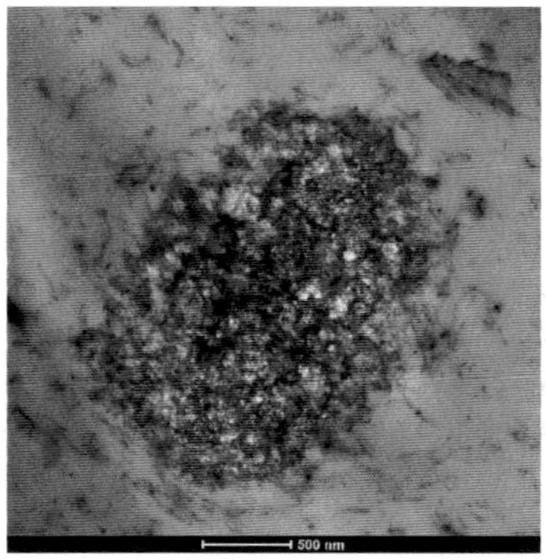

Figure 2. TEM image of composite containing 4.5 wt. % of GNP and 1.5 wt. % of MWCNT. The agglomeration is observed in between well-dispersed nanoparticles

The inner structure of the composites was investigated by means of transmission electronic microscopy. The microscopic images show that the most of filler particles are dispersed well and separated from each other (figure 1), but besides the separated particles all composites contain agglomerates (figure 2). As reported before [11, 12], the percolation threshold for the investigated composites was between 1.5–3 wt. % for MWCNT-filled, 3–6 wt. % for GNP-based and 1.5–3 wt. % for 50/50 bifiller composite. The percolation threshold for GNP particles in different polymers starts from ~5–10 % [3, 13].

The electromagnetic response of all investigated composites was measured in two microwave ranges of 0.01–18 GHz and 26–37 GHz (Ka-band) using MICRAN R4M vector analyzer and Elmika R2-408R scalar network analyzer respectively. The plane-parallel layers of investigated composites were precisely cut to match the dimensions of coaxial airline (in case of vector analyzer) and rectangular waveguide (scalar analyzer) and positioned normally to the wave vector of initial radiation (figure 3).

Figure 3. The sample fixture in rectangular waveguide (left) and coaxial airline (right).

The electromagnetic response was registered as ratios between the amplitude of reflected (transmitted) and incident radiation (S-parameters). The conductivity and complex dielectric permittivity were recalculated from S-parameters by standard methods [14] solving the following equations:

$$S_{11} = \frac{-j[(k_z/k_{2z})^2 - 1]\sin(k_{2z}\tau)}{2j(k_z/k_{2z})\cos(k_{2z}\tau) + [(k_z/k_{2z})^2 + 1]\sin(k_{2z}\tau)} \qquad (5)$$

$$S_{21} = \frac{2(k_{2z}/k_z)}{-2(k_{2z}/k_z)\cos(k_{2z}\tau) + j[(k_{2z}/k_z)^2 + 1]\sin(k_{2z}\tau)} \tag{6}$$

where τ is sample thickness, a is waveguide width, $k_z = \frac{\pi}{2\lambda}\sqrt{4a^2 - \lambda^2}$ and $k_{2z} = \frac{\pi}{2\lambda}\sqrt{4a^2\varepsilon - \lambda^2}$ ($k_z = \frac{2\pi}{\lambda}$ and $k_{2z} = \frac{2\pi\varepsilon}{\lambda}$), where λ is wavelength and ε is dielectric permittivity, are wavenumbers in waveguide and freespace (or coaxial airline) respectively.

4. Results & Discussion

4.1. Monofiller composites

The electromagnetic response of composites containing 0, 1.5, 3, and 6 wt. % of GNP and MWCNT was measured in both available frequency ranges. In both ranges MWCNT-based composites demonstrate higher dielectric permittivity and conductivity values than GNP-based ones.

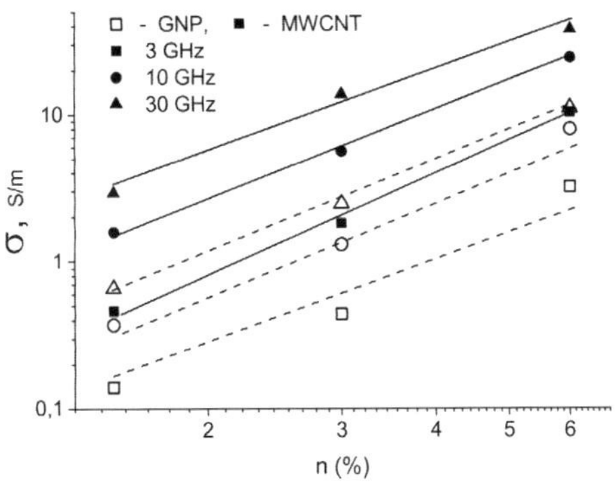

Figure 4. Composites conductivity dependence on filler content has exponential character similarly to a well-known empiric law $\sigma \simeq (n - n_0)^t$, where n_0 is percolation threshold concentration, $t \in [0, 1]$.

Such a behavior is in a good agreement with literature and can be explained by the difference in fillers aspect ratios [15] and volume resistivities (about 240 and <0.15 Ω·cm for GNP particles versus 1000 and >0.01 Ω·cm for carbon nanotubes). As shown at figure 4, the conductivity dependence on filler concentration follows the exponential law.

4.2. Bifiller effect to complex permittivity

Filler combinations with total concentrations of 3 and 6 wt. % were prepared. Their complex permittivity is in figures 5–6. The first to be noticed that the value of permittivity's imaginary part ε'' is proportional to MWCNT concentration for both tested filler contents. The highest permittivity values at the series with 3 wt. % of total nanocarbon concentration were reached by the composite with 3 wt. % of MWCNT. In contrary, for the series with 6 wt. % of total filler content the highest values of ε' in Ka-band were reached by the composites containing mixtures of two fillers (figure 6). In the lower frequency range the ε' values for MWCNT-only and 4.5 % MWCNT+1.5 %GNP composites are also very close.

Such a difference between 3 and 6 wt. % series can be explained by the fact that in first case the GNP concentration is lower than its percolation threshold, whilst in second case it is close to percolation. In case of 4.5 % MWCNT+1.5 %GNP the nanotubes exceed can compensate the above-percolation value of GNP content.

4.3. The synergistic effect observation

The permittivity additions for composites containing 1.5 and 3 wt.% of GNP, 1.5 and 3 wt.% of MWCNT and mixtures of total 3 and 6 wt.% (1.5 wt.% of GNP + 1.5 wt.% of MWCNT

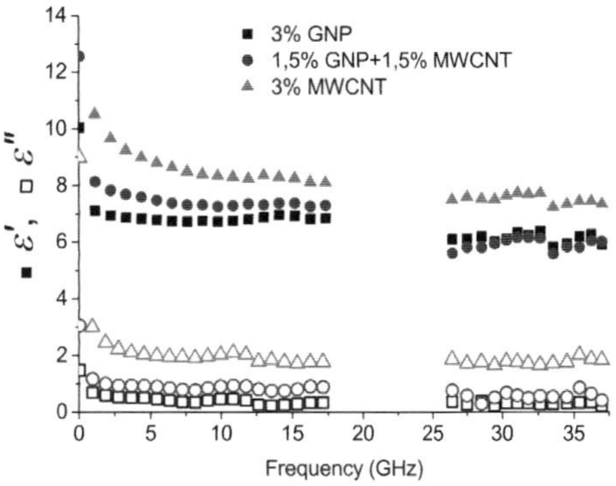

Figure 5. The complex permittivity of composites with 3 wt. % total filler content.

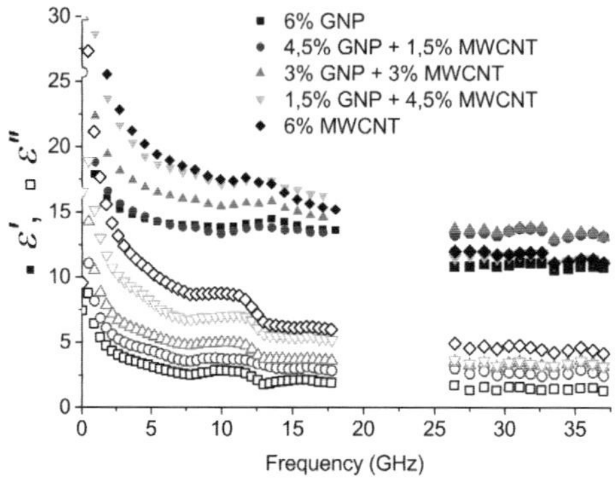

Figure 6. The complex permittivity of composites with 6 wt. % total filler content.

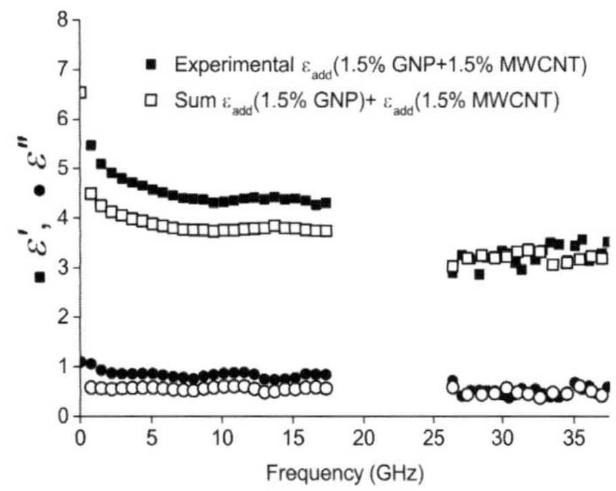

Figure 7. The complex permittivity addition by 1.5% GNP +1.5% MWCNT filler mixture vs sum of permittivity additions by 1.5% GNP and 1.5% MWCNT fillers. Experimental data is in good agreement with sum in 26–37.5 GHz range

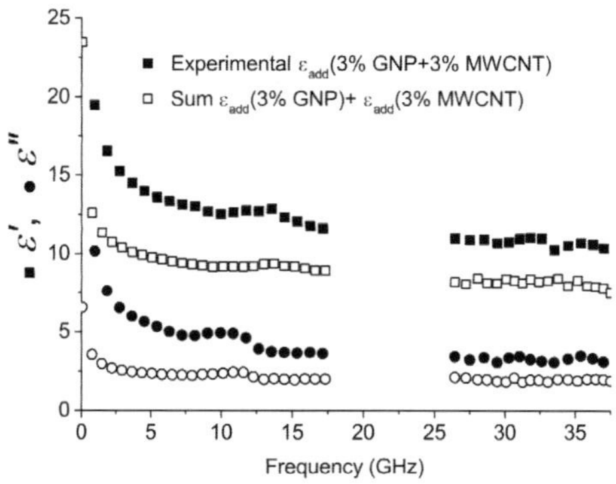

Figure 8. The complex permittivity addition by 3% GNP +3% MWCNT filler mixture vs sum of permittivity additions by 3% GNP and 3% MWCNT fillers. The value of separate filler additions is lower than experimental data

and 3 wt.% of GNP + 3 wt.% of MWCNT respectively) of nanocarbon content were calculated according equation 3. Monofiller additions were summarized and compared with bifiller ones.

According to obtained experimental data, for composites containing 3 wt.% of total nanocarbon content the synergistic exceed is more or less noticeable only in lower of two measured frequency ranges (figure 7). In Ka-band the electromagnetic response of bifiller composite is in agreement with a sum of two additions taken from monofiller composites. Thus in 26–37 GHz range the GNP and MWCNT fillers can be considered as non-interacting.

The different situation is observed at total 6 wt.% series (figure 8). The complex permittivity value increase of synergistic origin is noticeable in both measured ranges. Such an effect can be explained by interactions between fillers nanoparticles. MWCNT/GNP agglomerations must have different typical dimensions in comparison with aggregates of only GNP or MWCNT.

5. Conclusions

The PLA-based composites containing different concentrations of GNP, MWCNT and their combination were produced by melt-mixing technique and investigated in microwave frequency region. The frequency-dependent synergistic effect between GNP and MWCNT fillers was observed in form of a complex permittivity value increase. This effect makes possible the minimization of filler content which is vital for 3D-printing applications. The difference in effect of investigated fillers to real and imaginary parts of permittivity was demonstrated: MWCNT filler affects both parts of complex permittivity whilst GNP impacts mainly ε'. These fillers properties can be useful for the electromagnetic response tailoring which simplifies the fabrication of FDM filaments with specified dielectric permittivity. Such materials can be applied for the production of microwave optical elements, complex geometry electromagnetic shields, etc. The use of conventional melt-mixing method for composite preparation appears to be prospective: in spite of having a certain amount of agglomerates the most of filler particles are dispersed well. Further work in this direction must be focused on decrease of filler content and enhancement of dispersion stability.

Acknowledgements
This work is supported by H2020 RISE 734164 Graphene 3D project.

Polina Kuzhir and Dzmitry Bychanok are thankful for support by Tomsk State University Competitiveness Improvement Program.

References
[1] Gojny F H, Nastalczyk J, Roslaniec Z and Schulte K 2003 *Chemical Physics Letters* **370** 820–824
[2] Micheli D, Vricella A, Pastore R and Marchetti M 2014 *Carbon* **77** 756–774
[3] A Al-Hartomy O, A Al-Ghamdi A, Al-Salamy F, Dishovsky N, Shtarkova R, Iliev V and El-Tantawy F 2012 *International Journal of Materials and Chemistry* **2** 116–122
[4] Al-Hartomy O A, Al-Ghamdi A, Dishovsky N, Shtarkova R, Iliev V, Mutlay I and El-Tantawy F 2012 *Materials Sciences and Applications* **03** 453–459
[5] Bychanok D, Angelova P, Paddubskaya A, Meisak D, Shashkova L, Demidenko M, Plyushch A, Ivanov E, Krastev R, Kotsilkova R, Ogrin F Y and Kuzhir P 2018 *Journal of Physics D: Applied Physics* **51** 145307
[6] Wang Z, Luo J and Zhao G 2014 *AIP Advances* **4** 017139
[7] Paddubskaya A, Valynets N, Kuzhir P, Batrakov K, Maksimenko S, Kotsilkova R, Velichkova H, Petrova I, Bir I, Kertsz K, Mrk G I, Horvth Z E and Bir L P 2016 *Journal of Applied Physics* **119** 135102
[8] Kotsilkova R, Ivanov E, Todorov P, Petrova I, Volynets N, Paddubskaya A, Kuzhir P, Uglov V, Bir I, Kertsz K, Mrk G I and Bir L P 2017 *Journal of Applied Physics* **121** 064105
[9] Jiang X and Drzal L T 2012 *Journal of Applied Polymer Science* **124** 525–535
[10] Sarto M S, D'Aloia A G, Tamburrano A and De Bellis G 2012 *IEEE Transactions on Electromagnetic Compatibility* **54** 17–27
[11] Egiziano L, Lamberti P, Spinelli G, Tucci V, Kotsilkova R, Tabakova S, Ivanov E, Silvestre C and Di Maio R 2018 (Ischia, Italy) p 020152
[12] Spinelli G, Patrizia Lamberti, Vincenzo Tucci, Radost Ivanova, Sonia Tabakova, Evgeni Ivanov, Rumiana Kotsilkova, Sossio Cimmino, Rosa Di Maio and Clara Silvestre 2018 *Composites Part B: Engineering*
[13] Kopyt P, Salski B, Zagrajek P, Janczak D, Sloma M, Jakubowska M, Olszewska-Placha M and Gwarek W 2016 *IEEE Transactions on Terahertz Science and Technology* **6** 480–490
[14] Baker-Jarvis J, Geyer R G and Domich P D 1992 *IEEE Transactions on Instrumentation and Measurement* **41** 646–652
[15] Verma P, Saini P and Choudhary V 2015 *Materials & Design* **88** 269–277

IOP Conf. Series: Materials Science and Engineering **503** (2019) 012011 doi:10.1088/1757-899X/503/1/012011

Wood-based nitrogen doped activated carbon for fuel cells

A Plavniece[1,2], G Dobele[1], A Volperts[1], A Zhurinsh[1] and I Kruusenberg[3]

[1] Latvian State Institute of Wood Chemistry, Riga, Latvia.
[2] Rudolfs Cimdins Riga Biomaterials Innovations and Development Centre of RTU, Institute of General Chemical Engineering, Faculty of Materials Science and Applied Chemistry, Riga Technical University, Riga, Latvia.
[3] Institute of Chemistry, University of Tartu, Tartu, Estonia

ance.plavniece@inbox.lv

Abstract. The current study is devoted to the nitrogen doped carbons for application as fuel cell oxygen reduction catalysts, where wood char and highly porous activated carbons were used as precursors. Conditions of nitrogen introduction and its content in the carbonaceous material, as well as porous structure characteristics, such as specific surface and pore size distribution, were studied. It was found that the efficiency of doping reaction depends on the elemental oxygen content in the activated carbon. Relationships between nitrogen content, porous structure characteristics and electrodes electrochemical properties are demonstrated.

1. Introduction

Nowadays energy consumption constantly increases and the development of effective and cheap electrochemical sources of power, such as fuel cells and electrochemical capacitors, is highly relevant.

Power specifications and minimal price of H_2-O_2 fuel cells are limited by the expensive platinum-based catalysts, therefore prices of platinum increase while their reserves dwindle. Thus, the development of catalysts aimed to replace platinum-group metals is the most important for fuel cell design [1].

The main direction in the development of non-platinum catalysts for oxygen reduction is the study of cheap porous carbonaceous materials, which can be obtained by the pyrolysis of polymers including renewable biomass. It is well known that nitrogen atoms in carbon materials determine the properties of the doped activated carbons, such as high electrochemical stability, hardness, electric resistance, etc., as well as their catalytic properties [1,2] - doped materials have better electrical conductivity compared to precursors, which leads to high electrocatalytic activity [3].

There are several ways for a synthesis of porous carbon materials with N-containing heteroatoms using various nitrogen-rich compounds, such as ammonia, carbamide, melamine, etc. [4]. Traditionally N-doped carbons are synthesized using ammonia at high temperatures, by pyrolysis of nitrogen-containing carbon precursors [5], or doping with nitrogen-containing doping agents [6].

The lack of sufficient knowledge on the doping of the carbon materials calls for the ongoing researches of properties and structure of modified carbon matrix.

2. Experimental

Activated carbons based on alder wood char (SIA "Fille", Latvia) were obtained using chemical activation method with NaOH, activator ratios to precursor K = 2 and 3 and activation temperatures 600° C, 650° C and 700° C in the stream of argon, washed with deionised water, HCl, and water again up to neutral pH, and then dried overnight at 105° C. The main variables in the experiment were activation temperature and activator/charcoal ratio (further denoted as K).

Nitrogen was introduced into samples using dicyandiamide (DCDA) solutions with mass ratio carbon material/DCDA 1:20 and 1:10 and consequent treatment at 800° C. Two DCDA solvents were compared – ethanol (Et) and dimethylformamide (DFA) (the rest of the samples).

Specific surface area (m^2/g), micropore and total pore volumes (mm^3/g), and pore width (nm) were determined from nitrogen adsorption isotherms using the Quantachrome Nova 4200e instrument.

The electrochemical measurements were performed with a standard three-electrode system using the rotating disk electrode (RDE) method. The undoped, doped and reference materials (nanotubes and Pt-C) suspensions in ethanol (1 mg/ml) containing 0.0625% of Tokuyama OH^- ionomer AS-4 were pipetted onto the GC disk electrodes. The catalyst loading for all materials was thus 0.1 mg/cm^2.Saturated calomel electrode was used as a reference electrode and Pt foil was used as a counter electrode. The software used for controlling the experiments was Nova 2.1.2) and the potential was applied with a potentiostat/galvanostat Autolab PGSTAT 128N (Metrohm Autolab P.V). Measurements were performed in 0.1 M KOH solution at room temperature (23 ± 1 °C) at rotation rate 1900 rpm. For the substrate material, GC disks (GC-20SS, Tokai Carbon) with a geometric area (A) of 0.2 cm^2 were pressed into a Teflon holder and the electrodes were polished and cleaned.

3. Results and discussion

To investigate the obtained samples pre-and post-doping analysis of the porous structure of materials was carried out. Table 1 shows that wood charcoal has low porosity. In the case of nitrogen doped non-activated wood charcoal (samples with index N) specific surface, total pore volume and micropore volume decreased almost twofold after the doping process. As a result of the destruction of the micropores walls and the formation of larger pores at 800 ° C (doping temperatures), the average pore volume increases.

Choice of solvent and ratio of DCDA addition (10 or 20) did not influence porous structure significantly. Results are similar for Et and DMF, thus the solvent with better ability to dissolve DCDA, namely DFA, was chosen for the further experiments.

Table 1. Comparison of the porous structure of wood char (WC) depending on solvent and doping temperature.

Sample	Specific surface, m^2/g	Total pore volume, mm^3/g	Micropore volume, mm^3/g	Pore width, nm
WC	279	226	105	1.40
WC- DFA, 10N	145	229	40	1.96
WC- DFA, 20N	101	113,8	42	1.75
WC- Et, 20N	111	109	48	1.91

Liquid nitrogen sorption experiments revealed that wood charcoal has low porosity, and all of the activated samples have highly developed specific surface, however, their structure and pore size distribution are related to activation rate and temperature. The volume of adsorbed gas and the slope of isotherms for activated carbons (AWC) obtained at lower temperatures and especially using activator ratio K = 2 is characteristic for microporous sorbents (Figure 1). From the shape of the isotherm (Figure1) it can be observed that the adsorption mechanism of AWC obtained in 700° C and activator ratio K = 3 is different, which indicates that both micro- and mesopores are present [7].

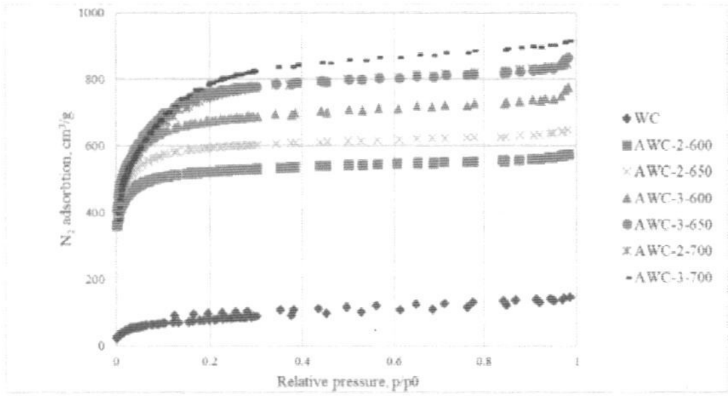

Figure 1. Nitrogen adsorption-desorption isotherms of wood char (WC), activated wood char (AWC) (activation temperature 600° C, 650° C and 700° C and activator ratio K = 2 and 3).

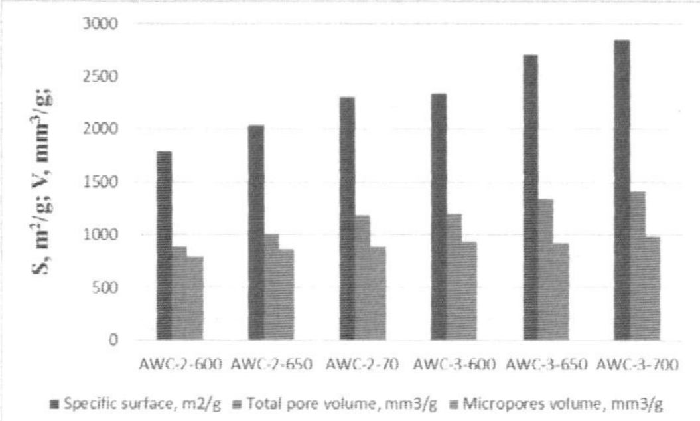

Figure 2. Comparison of the specific surface, total pore volume, and micropore volume of ACW depending on chemical activation conditions

Elemental analysis results are presented in Table 2 – it illustrates that nitrogen content is significantly influenced by aromatic structure defects and oxygen content of precursor. Non-activated wood char has the highest nitrogen content after doping – 20.67%. For all AWC samples, the nitrogen content is 5-6 %. Choice of DCDA solvent had a small influence on the nitrogen contents of doped wood char and AWC – the results are similar for ethanol and DFA.

Table 2. Elemental composition pre-and post-doping of wood char (WC) and activated wood char (AWC) (activation temperature 700° C, doping temperature 800° C)

Raw material	K	Solvent	Mass ratio carbon material/DCDA	N, %	C, %	H, %	S, %	O, %
WC	-	-	-	1.00	91.54	0.13	0.12	7.21
	-	DFA	10	19.13	75.13	0.18	0.15	4.41
	-	DFA	20	20.67	73.84	1.00	0.08	4.41
	-	Et	20	20.25	73.29	0.22	0.23	6.01
AWC	3	-	-	0.70	93.8	0.62	0.13	4.75
	3	DFA	20	5.55	90.97	0.89	0.12	2.48
	3	Et	20	6.17	89.60	0.14	0.09	4.00

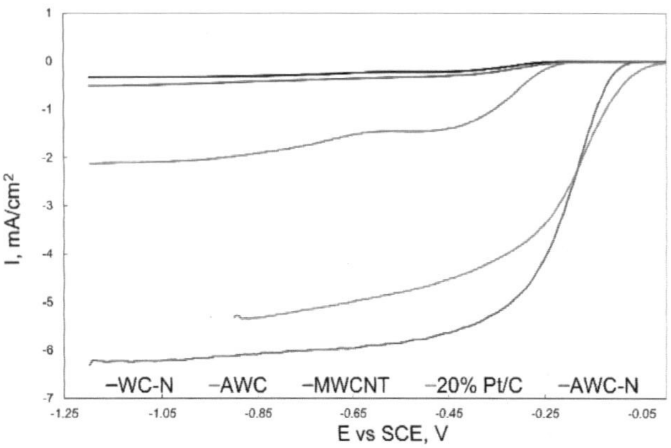

Figure 3. Comparison of RDE results for oxygen reduction on different catalysts in O_2 saturated 0.1 M KOH (v =10 mV/s, ω=1900 rpm)

For the tests of electrochemical properties sample of AWC, activated at 700° C, K=3 and subsequently doped at 800° C with DCMA, was refined and additionally treated at 800° C. Figure 3 illustrates catalyst polarization curves of the electrode made from this doped wood char (WC-N), activated wood char (AWC) and doped activated wood char (AWC-N) compared to commercial 20% Pt/C catalyst and carbon nanotubes (MWCNT) electrode. The shape of the polarization curves and diffusion current limits of WC-N, MWCNT demonstrate that they have low activity. The onset potential and half-wave potential for

AWC-N are much more positive compared to WC-N and MWCNT. The negligible negative shift of the onset potential is observable in comparison with 20% Pt/C, but at the same time diffusion limited current density is reaching much higher value in comparison to the commercial platinum catalyst. The overall electrocatalytic activity of doped AWC sample in KOH electrolyte was higher than that of commercial 20% Pt/C catalyst. The described electrocatalytic properties of the synthesized N-doped wood-derived carbon catalysts may be associated with the extremely high surface area and highly developed micro- and mesoporosity of the catalyst, as well as structural qualities of the biomass based carbon materials, e.g. presence of structural defects providing activity. It could also be associated with the high percentage of pyridinic nitrogen in the N-doped material and lack of stacking defects of graphene layers.

4. Conclusions

The studied samples of wood char were modified by introduction of nitrogen into their structure pre- and post- activation with the aim to apply them as catalysts for oxygen reduction in fuel cells. The most favourable doping reaction conditions, namely, doping agent dicyandiamide, DMF solvent and treatment temperature of 800 °C, were determined. It was found that the highest amount of nitrogen (more than 20%) can be introduced into wood char. The results show that nitrogen doped activated wood char (with 5-6% content of nitrogen) is a promising material for application in fuel cells for oxygen reduction.

Acknowledgements

The study was supported by M-ERA.NET project "Wood-based Carbon Catalysts for Low Temperature Fuel Cells (WoBaCat)" and LZP-2018/1-0194 project "Nanostructured Nitrogenated Carbon Materials as Promoters in Energy Harvesting and Storage Technologies (NN-CARMA)".

References

[1] Strelko V V, Kuts V S and Thrower P A 2000 On the mechanism of possible influence of heteroatoms of nitrogen, boron and phosphorus in a carbon matrix on the catalytic activity of carbons in electron transfer reactions *Carbon* **38** 1499 doi:10.1016/S0008-6223(00)00121-4

[2] Borghei M, Lehtonen J, Liu L and Rojas O J 2018 Advanced biomass-derived electrocatalysts for the oxygen reduction reaction *Adv. Mater.* **30** 1–27. doi:10.1002/adma.201703691

[3] Golberg D, Dorozhkin P S, Bando Y, Dong Z C,. Tang C C, Uemura Y, Grobert N, Reyes-Reyes M, Terrones H and Terrones M 2003 Structure, transport and field-emission properties of compound nanotubes: CNxvs. BNCx (x<0.1) *Appl. Phys. A Mater. Sci. Process.* **76** 499–507. doi:10.1I Kruusenber007/s00339-002-2047-7

[4] Boki K, Tanada S, Nobuhiro, O, Tsutsui S, Yamasaki R and Nakamura M 1987 Adsorption of polar and nonpolar gases of different sizes on nitrogen-containing activated carbon *J. Colloid Interface Sci.* **120** 286–8. doi:10.1016/0021-9797(87)90352-3

[5] Su F, Tian Z, Poh C K, Wang Z, Lim S H, Liu Z and Lin J 2010 Pt nanoparticles supported on nitrogen-doped porous carbon nanospheres as an electrocatalyst for fuel cells *Chem. Mater.* **22** 832–9. doi:10.1021/cm901542w

[6] Ratso S, Kruusenberg I, Vikkisk M, Joost U, Shulga E, Kink I, Kallio T and Tammeveski K 2014 Highly active nitrogen-doped few-layer graphene/carbon nanotube composite electrocatalyst for oxygen reduction reaction in alkaline media *Carbon* **73** 361–70. doi:10.1016/J.CARBON.2014.02.076

[7] Bansal R C and Goyal M 2005 *Activated carbon adsorption* Taylor & Francis

IOP Conf. Series: Materials Science and Engineering **503** (2019) 012012 doi:10.1088/1757-899X/503/1/012012

Electric field induced permittivity increase of carbon particle aggregates/silicone oil suspension

K Ozols, S Scegoleva and M Knite

Institute of Technical Physics, Faculty of Materials Science and Applied Chemistry, Riga Technical University, Paula Valdena 3/7, LV-1048, Riga, Latvia

E-mail: kaspars.ozols@rtu.lv

Abstract. Permittivity percolation threshold of carbon particle aggregates/silicone oil (CPA/SO) suspension was determined (Φ_C = 2.6 wt%). A direct current electric field induced permittivity increase versus time for the suspension containing 0.2 wt% of the CPA was measured at different field strengths. It was found that the electric field at 200 V/cm increases the permittivity of the CPA/SO suspension (ε' = 3.2 at 1 kHz) by three orders of magnitude in 30 min. At 800V/cm the same permittivity increase was reached in 1.5 min.

1. Introduction

Obtaining composites with desirable mechanical, thermal and electrical properties is a driving force for the research of polymer composites with aligned particle structure. Electroconductive particles/liquid dielectric matrix composites provide an opportunity for convenient studies of particle alignment dynamics [1,2] and electrical properties of the composites [1-4]. The alignment process can be accomplished using alternating current [1,2,5,6], direct current (DC) [3,4], and pulsed electric fields [7]. The influence of applied electric field strength [2-4], frequency [2,6], filler content [2-6] and matrix viscosity [2-4] has been determined. The main focus of the aligned liquid composite studies, overwhelmingly, is the electrical conductivity. In the present research, the DC electric field induced permittivity increase of the carbon particle aggregates/silicone oil (CPA/SO) suspension was studied.

2. Materials and methods

Medium viscosity (μ = 970 mPa·s) SO AK1000 (ε' = 2.75 at 1 kHz, $\sigma_{DC} = 2 \cdot 10^{-13}$ S/m) from Wacker Chemie AG was used as a matrix for the suspensions. Printex EX-2 carbon black (CB) (average particle size 30 nm, surface area 950 m^2/g, σ_{DC} = 370 S/m [8]) from Degussa AG was used as a filler. The suspensions were prepared as follows. The CB pellets were manually crushed into fine powder using pestle and mortar. The powder was gradually added to the SO while stirring. Subsequently, the filler was additionally dispersed using the pestle and, thus, the CPA/SO suspension was obtained. Suspensions containing from 0.1 wt% to 2 wt% of the CPA were prepared and used to determine a permittivity percolation threshold. The suspension containing 0.2 wt% of the CPA was used for all the alignment experiments. The CPA size in the suspension was estimated using size analyser Zetasizer Nano ZS90 from Malvern and Eclipse LV150 optical microscope from Nikon. It was found that the suspension contains 135 nm (peak) CPA and CPA agglomerates in sizes around 1-4 μm.

Permittivity measurements of the CPA/SO suspension were carried out in a custom-made measurement cell (Figure 1). The size of the copper electrodes is 25 mm x 40 mm. U-shaped 2 mm thick silicone rubber spacer was used to separate the electrodes while holding the suspension. The suspension was injected into the cell using a syringe. Measurements of a capacitance (measurement voltage 0.2 Vrms) and a DC conductance were accomplished using Agilent E4980A LCR meter. The LCR meter was also used as a source of the DC alignment voltage (for field strengths above 200 V/cm a separate power supply was used). A measurement sequence: electric field was applied for a chosen time interval, electric field was switched off for 30 s, data readings, the field switched on, etc. A flat polytetrafluoroethylene (PTFE) substrate equipped with 5 mm wide copper electrodes separated by 2 mm gap was used to image the aligning CPA network in the suspension layer through the microscope while, simultaneously, receiving readings from the LCR meter.

Figure 1. Components of the suspension permittivity measurement cell and the cell in the PTFE holder.

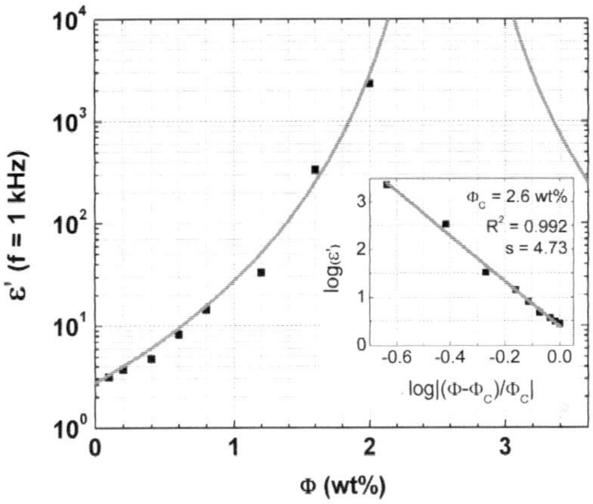

Figure 2. Permittivity vs. filler concentration for unaligned CPA/SO suspensions. The continuous line is a theoretical fit obtained using Eq. (1).

3. Results and discussion

In order to choose the CPA/SO suspension with a potentially large permittivity change during the alignment, a permittivity percolation threshold was determined. The relative permittivity (at 1 kHz) versus CPA filler concentration for unaligned suspensions is shown in Figure 2. From the work of Dubrov et al [9] it follows that the permittivity can be calculated using equation:

$$\varepsilon' = \varepsilon'_1 \left| \frac{\Phi - \Phi_C}{\Phi_C} \right|^{-s} \quad (1)$$

where ε' is the relative permittivity of the composite, ε'_1 is the relative permittivity of the matrix, Φ is the filler concentration, Φ_C is the critical filler concentration (the percolation threshold), s is the critical exponent. The results for the permittivity measurements were found using a linear fit for log(ε') versus $\log|(\Phi-\Phi_C)/\Phi|$ (Figure 2, the insert). The best fit was obtained at $\Phi_C = 2.6$ wt% (s = 4.73). In the case of a DC conductivity, previously determined Φ_C value [4] for the suspension was 0.6 wt% (exponent 11.48). Thus, the Φ_C for the conductivity is significantly lower than for the permittivity. As pointed by Grady [10], percolation thresholds for carbon nanotube/polymer composites, which are calculated from the dielectric constant and conductivity, are identical or nearly identical. As mentioned by Knaapila et al [6], percolation depends on particle topology and is also determined by the interactions among particles and between particles and matrix. The dissimilar percolation threshold values for the suspension suggest that, although the filler at small concentrations quickly forms conductive channels with tunneling conductivity, these channels are long and sparse and, unlike it is in case of solid composites, they form a smaller series connection capacitance and provide a small contribution to the permittivity.

The electric field induced relative permittivity (1 kHz) and a DC conductivity change for the CPA/SO suspension is shown in Figures 3 and 4, respectively. Application of the field increases the permittivity by up to almost 3 orders of magnitude in 1.5 min (for 800 V/cm). At the same time, the DC conductivity of the suspension increases by over 4 orders of magnitude.

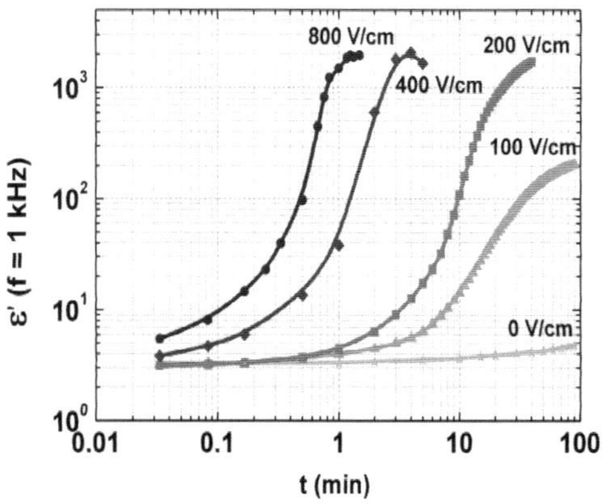

Figure 3. The relative permittivity (at 1 kHz) versus time for the CPA/SO suspension containing 0.2 wt% of the CPA depending on the DC field strength.

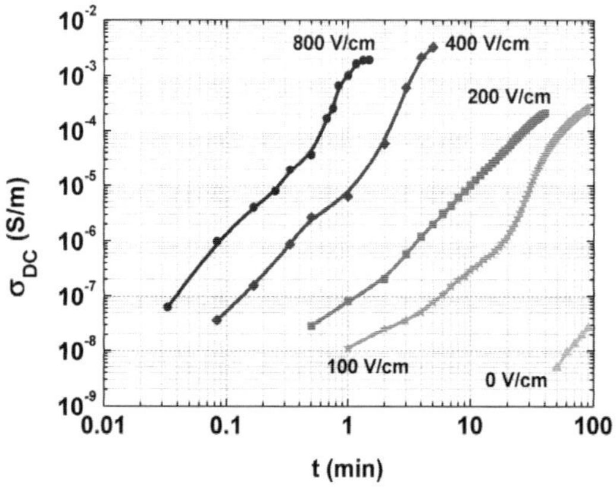

Figure 4. The DC conductivity versus time for the CPA/SO suspension containing 0.2 wt% of the CPA depending on the DC field strength.

During repeated experiments, it was observed that a large (up to 1 order of magnitude or more) dispersion for the ε' and the σ_{DC} saturation values was obtained (not seen in the Figures due to averaging) for each field strength. In addition, it was often observed that the permittivity decreased after reaching the saturation value (as seen in 400 V/cm graph). This effect, most likely, is caused by two factors: 1) expansion of the SO matrix due to significant Joule heating (up to 50^{0}C) at the end of the alignment; 2) over-densification of the particle agglomerate network, which excludes a space for micro-capacitors formed by nearby particles. Here, it is important to note that if the field was off for about 10 min, the permittivity increased by up to 15% (further lack of the aligning field led to suspension permittivity decrease).

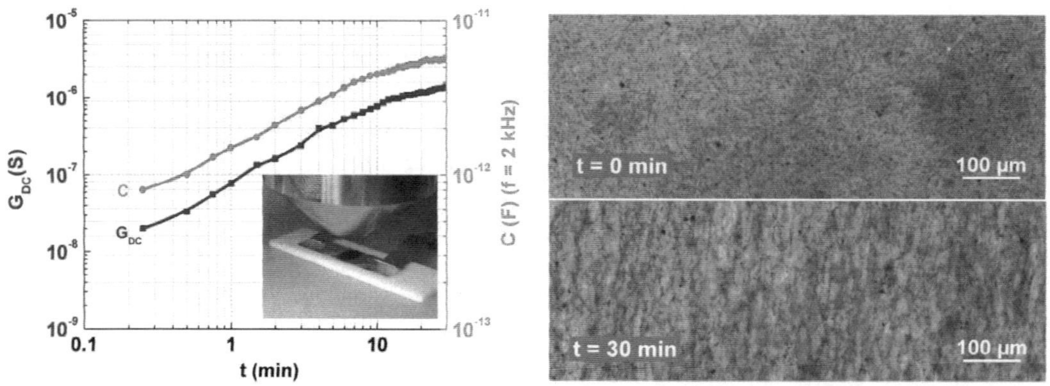

Figure 5. Alignment (200 V/cm) of a thin (~ 0.2 mm) layer of the CPA/SO suspension containing 0.2wt% of the CPA on the PTFE substrate with simultaneous capacitance and conductance measurements while taking snapshots of the aligning particle network through the microscope. The field direction on the micrograph is vertical.

To follow a correlation between the emerging aligned CPA network and the change in the electrical properties, micrographs of the suspension were taken during the alignment while performing measurements of the capacitance and the DC conductance (Figure 5), simultaneously. It can be seen that the capacitance curve is closely related to the conductance curve although the C increase is around one 1 order of magnitude, but the G_{DC} increase is around 2 orders of magnitude. The micrographs show that the CPA agglomerates reallocate forming chains, which are predominantly aligned along the electric field lines (the CPA particle chaining process is thought to be similar as described by Oliva-Avilés et al [1]). In terms of an equivalent circuit, series (and to less extent parallel) connection capacitor-resistor chains are formed causing the observed permittivity and conductivity increase of the suspension.

4. Conclusions

Application of the DC electric field (200 V/cm and higher) leads to the permittivity increase of the CPA/SO suspension (containing 0.2wt% of the CPA) by up to three orders of magnitude. At lower field strength (100 V/cm) the permittivity increase reduces to two orders of magnitude. The increase of the permittivity is caused by the alignment of initially randomly positioned CPA and their agglomerates, which form a network of a branched structure of microcapacitor chains, which densify upon increasing electric field strength and time.

5. Acknowledgements

This research was supported by the Latvian National Research Programme in Materials Science (IMIS2). The authors also would like to thank "IMCD Baltics UAB" for supplying the silicone oil.

References

[1] Oliva-Avilés A I, Alonzo-García A, Zozulya V V, Gamboa F, Cob J and Avilés F 2018 A dielectrophoretic study of the carbon nanotube chaining process and its dependence on the local electric fields *Meccanica* **53** 2773-91

[2] Oliva-Avilés A I, Avilés F, Sosa V, Oliva A I and Gamboa F 2012 Dynamics of carbon nanotube alignment by electric fields *Nanotechnology* **23** 465710

[3] Badard M, Combessis A, Allais A and Flandin L 2016 Electric field as a tuning key to process carbon nanotube suspensions with controlled conductivity *Polymer* **82** 198-205

[4] Knite M, Linarts A, Ozols K, Tupureina V, Stalte I and Lapcinskis L 2017 A study of electric field-induced conductive aligned network formation in high structure carbon black/silicone oil fluids *Colloids and Surfaces A: Phys. Eng. Asp.* **526** 8-13

[5] Yakovenko O, Matzui L, Danylova G, Zadorozhnii V, Vovchenko L, Perets Y and Lazarenko O 2017 Electrical Properties of Composite Materials with Electric Field-Assisted Alignment of Nanocarbon Fillers *Nanoscale Research Letters* **12**:471

[6] Knaapila M, Høyer H, and Helgesen G 2014 Dielectrophoresis in particle confinement: Aligned carbon particles in polymer matrix below percolation threshold *Eur. Phys. J. Special Topics* **223** 1869–82

[7] Huynh M T T, Cho H B, Suzuki T, Suematsu H, Nguyen S T, Niihara K and Nakayama T 2018 Electrical property enhancement by controlled percolation structure of carbon black in polymer-based nanocomposites via nanosecond pulsed electric field *Compos. Sci. Technol.* **154** 165-74

[8] Pantea D, Darmstadt H, Kaliaguine S and Roy C 2003 Electrical conductivity of conductive carbon blacks: influence of surface chemistry and topology *Appl. Surf. Sci.* **217** 181–93

[9] Dubrov V E, Levinshtein M E and Shur M S 1976 Permittivity anomaly in metal-dielectric transitions. Theory and simulation *Zh. Eksp. Teor. Fiz.* **70** 2014-24

[10] Grady B P 2011 Carbon nanotube-polymer composites: manufacture, properties, and applications John Wiley & Sons Inc ISBN: 9781118084380 268–9

IOP Conf. Series: Materials Science and Engineering **503** (2019) 012013 doi:10.1088/1757-899X/503/1/012013

Synthesis and studying of reduced few-layered graphene coatings in gas sensor applications

Ingars Lukosevics, Peteris Lesnicenoks and Janis Kleperis

Institute of Solid State Physics, University of Latvia, 8 Kengaraga str., LV-1064 Riga, Latvia

ingars@cfi.lu.lv

Abstract. In this work reduced few-layered graphene (rFLG) nanoparticles were synthesized using electrochemical pulse exfoliation method from waste graphite crucibles. The regular change in voltage polarity in the synthesis process ensures both the separation of graphite in layers and the reduction of graphene oxide. A method for synthesizing free-standing rFLG and nitrocellulose (NC) composite film has been developed involving creation of rFLG-NC ink that can be deposited on various substrates. It has been observed that a successful synthesis of a free-standing composite coating is possible with the mass ratio of rFLG:NC at least 9:1 of which resistivity is on the order of approximate 10 ohm-centimetres in dependence from temperature and surrounding atmosphere. Sensor electrode was prepared by simple dip-coating method and the response to humidity, organic solvent vapours and nitrogen containing gases was measured and conclusion made that rFLG-NC coating is sensitive to many kinds of gases (cross-sensitive) what may result in false detecting or can be used to multi-sensor chips for artificial olfaction devices.

1. Introduction

One of the biggest sources of energy at present day is through the fossil fuels. Burning of which releases dangerous side products. The most dangerous are considered to be nitrogen oxides NO and NO_2. For successful limitation of these pollutants gas sensors are required. A potential gas sensing material is carbon allotrope graphene [1,2]. Many methods have been developed other than first used micromechanical cleavage [2]. For preparation of relatively large size graphene monolayers chemical vapour deposition is widely used but this method is difficult to scale to industrial levels [3]. Other than graphite exfoliation through oxidation of graphite [4] electrochemical exfoliation is a potential method for scalable synthesis of graphene sheet stacks [5,6]. For even more effective exfoliation usually graphene sheet stack containing suspension is additionally subjected to ultrasonication [7].

For further applications variety of graphene-based materials are prepared ranging from thin gas permeable membranes [8] to three dimensional structures for energy storage [9].

One option for creation of larger structures is by creation of graphene – polymer composite through either *in situ* polymerisation or through dissolution [10]. In addition to create a relatively macroscopic material, some composite material properties can be tailored enhancing for example mechanical [11] or electrical [12] properties.

Gas sensors are classified by their detection mechanism, one of which is the chemical reaction or adsorption taking place in the surface of a semiconductor (e.g. tin dioxide is widely used [13]) in result of which the electrical resistance increases or decreases depending on the reaction taking into account for the donor or acceptor presence in the semiconductor [14]. Sensor response *S* in percentage is introduced to characterise this change of resistance and described elsewhere [14,15].

2. Experimental

2.1 Synthesis and reduction of rFLG

Graphene sheet stack (GSS) powder was synthesised through electrochemical pulse exfoliation of waste graphite crucible from bronze metallurgy as the working electrode in 1 M H_2SO_4 electrolyte, counter electrode was platinum. Potential to the cell was applied periodically, +10 V for 3 seconds and -10 V for 5 seconds. The resulting suspension was vacuum filtered with DI water and dried in an oven. Dry GSS powder was then mixed with N,N-Dimethylacetamide in a sealable glass jar, which was further placed in sonication bath for multiple hours. After sonication the GSS powder was again vacuum filtered with DI water and dried in an oven. For further reduction dried powder was put in an alumina tube furnace purged with Ar/H_2 95:5 gas mix at temperatures up to 800 °C.

2.2 Preparation of rFLG-NC films

Desired mass ratios between rFLG and NC was mixed in acetone and sonicated for around 3 hours to create the ink-like suspension.

For preparation of free-standing composite films aluminium foil was adhered to sample glass and used as a substrate. Ink was deposited on the aluminium foil layer by layer via tape casting in ambient conditions. To remove rFLG-NC film from substrate, it was slowly put into 0.5 M KOH solution with the aluminium side facing solution. Separated rFLG-NC film remained floating in the solution and was carefully taken out on PTFE strip and dried on a laboratory hot plate (around 50 °C) for an hour. Samples for gas and vapour sensing measurements were prepared by dip-coating method from the rFLG-NC-acetone suspension (rFLG:NC mass ratio of 4:1) on sample glass pieces (width 10 mm and length 15 mm).

2.3 Characterisation of rFLG-NC films

To find out the type of conductance of the prepared film, Hall effect measurement was carried out on HMS-5000 *van der Pauw* Hall effect measurement system.

The resistance as ohms per square was measured by placing two identical thin 5 mm long electrodes parallel to each other at a distance of 5 mm.

The microscopic images of the composite surface were obtained by the means of *Phenom Pro* desktop SEM with acceleration voltage of 10 kV.

2.4 Gas sensing measurements

Gas sensing measurements were carried out at room temperature. The prepared samples were placed in a sample holder and electrodes were pressed onto the sample film 2 mm apart. The holder was placed in a closed volume – desiccator. The electric measurements were carried out with potentiostat *VoltaLab PGZ 301*. Test substance was introduced, evaporated in a closed desiccator. To achieve complete evaporation of water and ammonia solution the desiccator was held at an elevated temperature of 64 °C. For tests with nitric oxide and nitrogen dioxide the gases were introduced by reacting copper metal with dilute and concentrated nitric acid respectively.

3 Results and discussion

3.1 Characterisation of composite films

The Hall effect measurement indicated that the conductance of synthesized rFLG-NC material is semiconducting p-type. The reason behind this is thought to be some oxygen containing functional groups on rFLG that most likely were introduced in the electrochemical exfoliation and persisted through reduction process. The resistivity dependence from the electrically conducting rFLG particle percentage in the composite (Figure 1) show an obvious trend of decreasing resistivity when the amount of rFLG in composite is increased. The film thicknesses in each of the rFLG percentages varies from 5 μm to 40 μm and are taken into account when calculating resistivity.

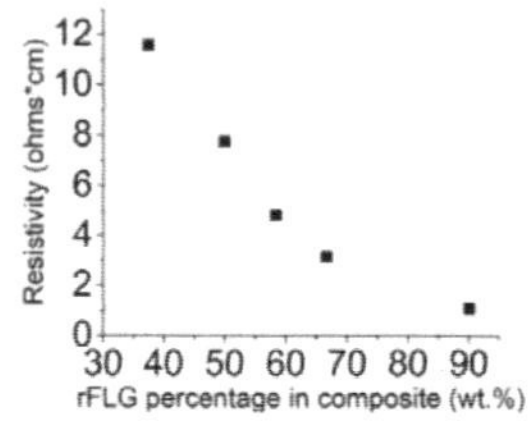

Figure 1. Composite resistivity vs the rFLG percentage in composite by mass.

The free-standing (Figure 2 (a)) and supported (Figure 2 (b)) films show distinct differences on their structures. The supported film has rFLG sheets assembled more planar to each other in comparison to the free-standing film likely due to the synthesis of each type of film. Utilization of relatively large amounts of ink in combination of quickly volatile solvent used in blade casting technique did not let the particles to align themselves with the least fill factor *i.e.* parallel to each other as it was observed in the dip-coating method where it is intuitively understandable that each planar rFLG particle would align as parallel as possible to the substrate surface.

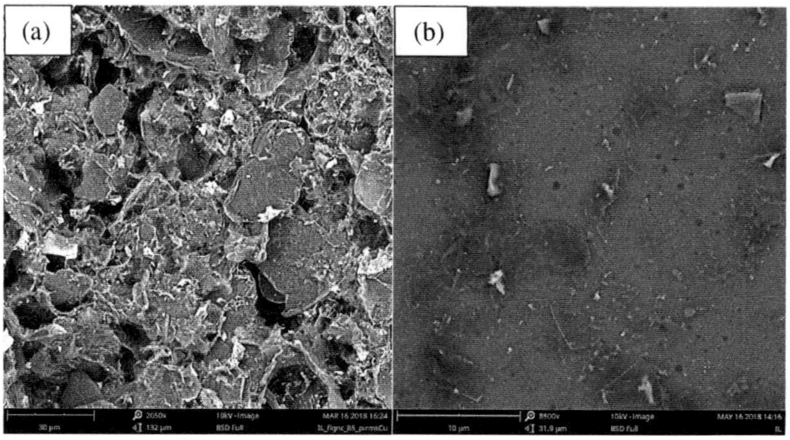

Figure 2. SEM images of two types of the made films. Scale lengths indicate 30 μm (a) and 10 μm (b).

3.2 Gas sensing measurements

Resistance over time of the film was obtained by applying a constant potential across the sample so that current would be on the order of few μA. This was done to make the resistance change due to Joule heating negligible.

Advantage or disadvantage in some cases of using composite material as the sensor material is that resistivity can change due to each part of the component of composite. Figure 3 shows the resistance change in interaction with solvent vapours – acetone, methanol and acetonitrile – each of which is known to dissolve polymer NC. The reversibility of resistance change in five or ten minutes is understandable, considering the volatile nature of solvents in ambient conditions. The reversibility also indicates that no structural changes are happening *i.e.* rFLG particles are not noticeably changing their distribution in the film. In addition, consecutive measurements under the same conditions in solvent vapours were carried out five times and these measurements did not show any significant

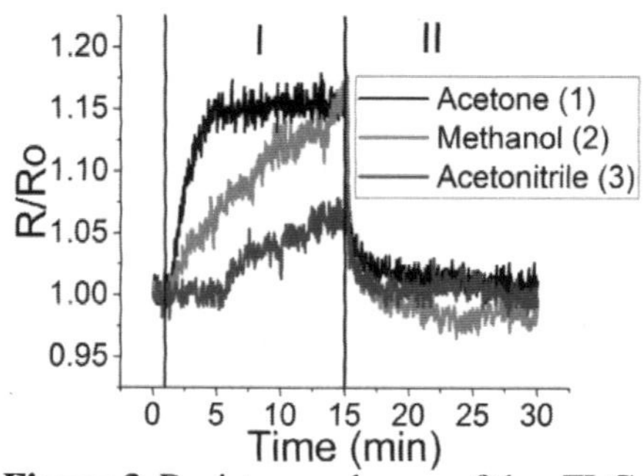

Figure 3. Resistance change of the rFLG-NC supported film under the influence of acetone, methanol, acetonitrile vapours. Part I shows the evaporation of target substance and the sensing reaction of film and II shows the purge of the closed volume.

difference or trend. It must be pointed out that for acetone and methanol sensors part of the contribution could be from chemical reactions between target gas and oxygen derivatives O^- adsorbed to rFLG surfaces [16,17]. The electrons generated by the reaction could recombine with the main charge carrier type (holes) in the composite and the overall resistance would increase. In the case for acetonitrile most likely there is no reaction happening and the response is purely due to interactions between the vapours and the nonconductive part of composite – NC. Although a sensing mechanism it must be pointed out that the resistance change to these organic solvents is relatively small in relatively large concentrations of solvent vapours and this reaction must be considered if composite is used as sensing material for other substances e.g. ammonia or nitrogen oxides described further.

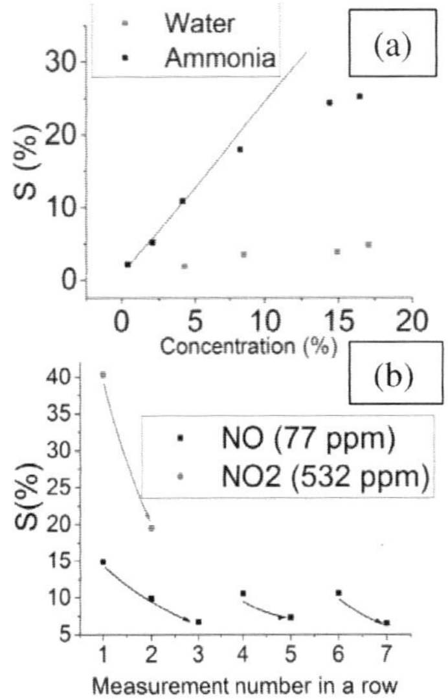

Figure 4. Response vs concentration for ammonia and water vapour (a) and response vs measurement number for nitrogen oxides showing the reversibility of sensor film, arrows indicate consecutive measurements (b).

Ammonia in water solution was evaporated in the closed container and Figure 4 (a) shows the response values vs concentration to both ammonia and humidity. In contrast to exposure to nitrogen oxides where the resistance decreased the resistance due to ammonia increased. These observations agree with the fact that composite film is a p-type semiconductor. Ammonia a less electronegative compound adsorbs to rFLG sheets and donates electrons that recombine with holes in rFLG thus increasing the overall resistance [14]. Whilst the more electronegative nitrogen oxides when adsorbed to rFLG attract the electrons increasing the hole concentration in material and decreasing the overall resistance. It is important to note that the resistance change in exposure to nitrogen oxides is not reversible in short periods of time (15 minutes) as seen from Figure 4 (b). Reaction mechanism of nitrogen oxides is definitely adsorption to rFLG considering that there would be no possible reaction between the NC as it is fully nitrated, no more place available for nitrite ions to settle.

Conclusions

We have synthesised rFLG through electrochemical exfoliation of waste graphite crucible and incorporated it into NC matrix in the form of free-standing and supported films. The supported films show a more planar orientation of rFLG sheets than in the case of freestanding. The composite film shows p-type conductance both from Hall and gas sensing measurements. We have shown that utilisation of composite material for sensor material can provide response reaction from both parts of the composite – NC solvent vapours and gases interacting with rFLG NH_3, NO_2 and NO. Composite film is able to detect nitric oxide at concentrations as low as 77 ppm but the film electrical resistance in relatively short periods of time is partly non-reversible not as in the case of solvents, water and ammonia.

Acknowledgements

Authors acknowledge support from Latvian Council of Science Project No 2018-1/0194.

References

[1] Schedin F, Geim A K, Morozov S V, Hill E W, Blake P, Katsnelson M I and Novoselov K S 2007 *Nat. Mater.* **306** 652–5
[2] Novoselov K S, Geim A K, Morozov S V, Jiang D, Zhang Y, Dubonos S V, Grigorieva I V and Firsov A 2004 *Science* **306** 666–9
[3] Chen X, Zhang L and Chen S 2015 *Synth. Met.* **210** 95–108
[4] Rae C 2018 *Energy Storage Mater.* **14** 8–21

[5] Liu J and Theses S 2017 *Graphene-based composites for electrochemical energy storage* (Springer) pp 39–51
[6] Lesnicenoks P, Berzina A, Lukosevics I and Grinberga L 2017 *Proc. Est. Acad. Sci.* **64** 403–8
[7] Qi X, Zhou T and Deng S 2014 *J. Mater. Sci.* **49** 1785–93
[8] Liu G, Jin W and Nanping X 2015 Graphene-based membranes *Chem. Soc. Rev.* **44** 5016–30
[9] Song Y, Liu T, Qian F, Zhu C, Yao B, Duoss E, Spadaccini C, Worsley M and Li Y 2018 *J. Colloid Interface Sci.* **509** 529–45
[10] Du J and Cheng H-M 2012 *Macromol. Chem. Phys.* **213** 1060–77
[11] Yang X, Shang S and Liang L 2011 *J. Appl. Polym. Sci.* **120** 1355–60
[12] Li Y, Samad Y A, Polychronopoulou K, Alhassan S M and Liao K 2015 Highly *Sci. Rep.* **4** 4652
[13] Andio M A, Browning P N, Morris P A and Akbar S A 2012 *Sensors Actuators, B Chem.* **165** 13–8
[14] Soleimanpour A M, Jayatissa A H and Sumanasekera G 2013 *Appl. Surf. Sci.* **276** 291–7
[15] Wu Z, Chen X, Zhu S, Zhou Z, Yao Y, Quan W and Liu B 2013 *Sensors Actuators, B Chem.* **178** 485–93
[16] Liu F, Chu X, Dong Y, Zhang W, Sun W and Shen L 2013 *Sensors Actuators, B Chem.* **188** 469–74
[17] Fong C F, Dai C L and Wu C C 2015 *Sensors (Switzerland)* **15** 27047–59

IOP Conf. Series: Materials Science and Engineering **503** (2019) 012014 doi:10.1088/1757-899X/503/1/012014

Specific features of magnetic domain structure in epitaxial magneto-optical ferrite-garnet films

A I Ivanova, E M Semenova and M K Virchenko

Tver State University, 170100 Tver, Russia

e-mail: alex.ivanova33@yandex.ru

Abstract. We describe the experimental studies of both domain and crystal defect structure of magneto-optical bismuth-substituted ferrite–garnet films (FGF) with high Faraday effect. Indentation and scratching (sclerometric) tests of either magnetically uniaxial or easy plane anisotropy ("planar") films were carried out in parallel with traditional chemical etching. We demonstrate the effect of strong local changes of magnetic anisotropy on the domain structure due to local mechanical stresses. New types of biperiodic domain structure imposed by external stresses were found to exist.

1. Introduction

The interest in epitaxial Bi-substituted ferrite-garnet films $Bi_xGd_yLu_{3-x-y}Ga_zFe_{5-z}O_{12}$ (Bi:FGF), thickness 3–7 μm, deposited on oriented $Gd_3Ga_5O_{12}$ substrates is related to the possibility of using them in various devices (visualizators of spatial magnetic field microdistributions, optical switches, modulators, isolators, *etc.*). Among the advantages of FGF are high resolution and contrast imaging of the magnetic field distributions associated with large values of Faraday rotation. To achieve such values, Bi ions are introduced into the crystallographic lattice of the FGF. The domain structure (DS) of uniaxial FGF was investigated in great detail by many researchers. [1-4]. At the same time, however, the studies of planar DS till now have received but scant attention.

In this paper we examine both uniaxial films with the easy axis direction normal to the plane of the film having a "labyrinth" domain structure, and planar films with magnetic moment oriented in the *plane* of the film. Domain structure of magnetic films is determined by the balance of many factors, of which the exchange interaction, magnetization, anisotropy, the shape of a magnetic sample, the energy of domain walls, the magnitude and direction of external magnetic field, temperature, defects, magnetostriction, the surface structure. Defects of the GGG substrate and films which arise during the growth and processing, various kinds of surface disturbances affect the magnetic characteristics of FGF. Generally high-quality FGF should possess defect density as low as possible, so the study of defects, both structural and mechanically induced, is an important research task.

We study the intrinsic (DS) of "planar" FGF. For this film type the magnetocrystalline anisotropy E_a is small compared with the shape anisotropy $E_s = \mu_0 M_s^2 \sin^2\varphi$, where M_s is saturation magnetization and φ is the angle between the film plane and $\mathbf{M_s}$ The total energy $E = E_s + E_H = K_s \sin^2\varphi - H_{ext} M_s \cos(\theta - \varphi)$, where θ is the angle between $\mathbf{H_{ext}}$ and film plane. For $\theta = \pi/2$, the magnetization z-component M_z is a linear function of the external field and $M_z \leq M_s$ [5]. This property of planar FGF favours their use as magneto-optical sensors visualizing the microfield. Both uniaxial and planar FGF are finding their use for this purpose. Here it should be stressed, however, that in these both cases it is normal field component that is solely applied for sensing. Moreover, in standard microscope arrangement the intrinsic DS of planar FGR remains invisible due to Faraday effect symmetry. In the present work we identify the intrinsic DS of planar FGF and observe directly the processes of domain wall displacement. To this end we developed a manipulation tool enabling to orient the sample at the microscope table at any desired angle. Optimum is chosen on the basis of negotiation between the Faraday contrast and image distortion due to sample inclination. In this way we were able to prove that FGF under study possesses well developed structure of 90- and 180-degree domains (Figure 1). By direct observation evidence was obtained of the smallness of the "planar" coercive field, H_c, and Barkhausen jumps typical of domain wall displacement processes. This property will be useful for the improvement of development of high-sensitivity magnetometry.

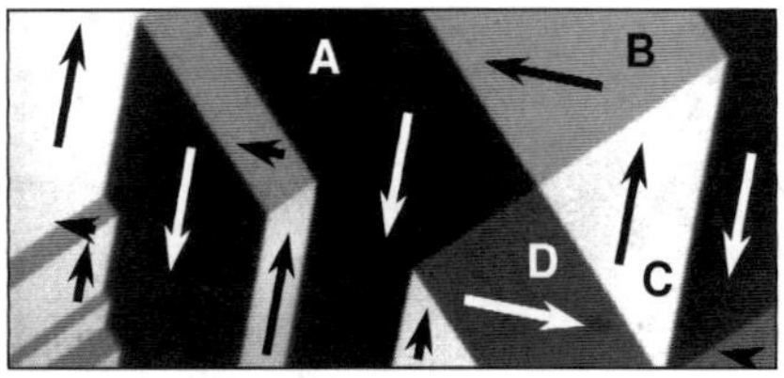

Figure 1. Domain structure of planar transparent FGF as revealed by Faraday effect

2. Experimental

Bi-FG films are transparent single-crystal layers grown by liquid-phase epitaxy on a non-magnetic crystalline $Gd_3Ga_5O_{12}$ substrates was obtained by Czochralski method. In the experiment in addition to domain structure we studied structural and mechanically induced defects in the film and in the GGG substrate. Defects in the substrate are formed during crystal growth (inclusions, growth and dislocation bands) and during processing (scratches, chips, etc.). Defects and heterogeneities of the substrate are inherited by the film, causing local changes in the magnetic properties. The dislocation structure of crystalline $Gd_3Ga_5O_{12}$ substrates, uniaxial and planar magnetic films was revealed by the method of selective chemical etching. The mechanically polished GGG substrate and magnetic films were etched in 85% phosphoric acid at temperature of 160°C and 100°C respectively. The scratching experiments were carried out on Vicker's hardness meter, dislocation structure was investigated by optical interference profilometry and scanning electron microscopy (SEM) with secondary electron (SEI) and back-scattered electrons (BSE) while the magnetic domains were observed by means of magneto-optical microscopy.

3. Results and discussion

The dislocation structure of GGG substrate and FG film grown in [111] was determined by selective chemical etching. Figure 2 (a, b) shows large (150-170 μm) etch pits of the jogged screw dislocations with a sharp bottom indicating the growth origin of the dislocation. The small (5-7μm) etch pits of the magnetic film correspond to the straight edge dislocations (Figure 2 c). Dimensions of dislocation pits and bottom form are determined by optical profilometry.

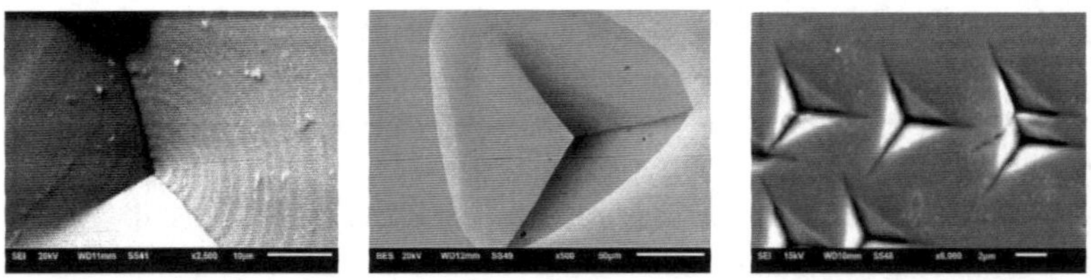

a *b* *c*

Figure 2. SEM image of dislocation etch pit: *a*) GGG (SEI), *b*) GGG (BSE), *c*) FGF surface (SEI).

Single dislocations do not have a noticeable effect on the magnetic properties of the film, but dislocation clusters cause formation of microdomains, shape distortion of domain walls and the change of domain sizes. Defects in the film, caused by non-conformities of parameters of the lattices of the substrate and of the film, lead to the formation of dislocations, rupture of the film and negatively affect the quality of thermomagnetic recording on the ferrite-garnet film.

The motivation for our research were studies of Jordi Sort [6-7] and Broese van Groenou [8] . Using sclerometry (scratching an edge of a quadrangular pyramid) with a load of 20 g we scratched on uniaxial films with a labyrinth domain structure. We observed a shift of domains in the defect region by the magneto-optic (MO) microscopy based on polar Kerr effect (Figure 2a). The external magnetic field (10 mT) applied in the direction normal to the film plane caused changes of configuration of domains and the inclusion of this defect in the domain of one sign (Figure 2b)

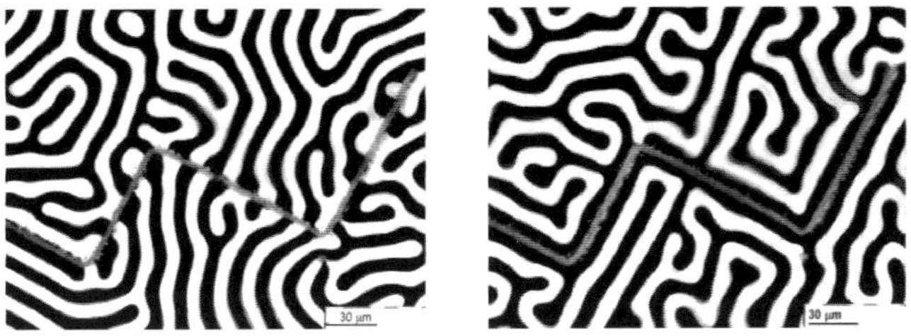

a *b*

Figure 3. MO image of scratch-modified domain structure:
a– zero external field, *b* – applied field 10 mT.

The effect of induced mechanical defects on the domain structure was also investigated on planar ferrite-garnet films. The indentation and scratch tests were carried out on Vicker's hardness metter with a load of 20 g and 50 g. The appearance of closure domains in the defect region was visualized (Figure 4a). Under the action of a NdFeB permanent magnet (10 mT) along the film plane and normal to it, we observed a partial restructuring of closure domains and the alignment of part of the domains along the stress fields. On the scratch in the form of "zigzag", the closure domains with their own fine domain structure were visualized (Figure 4b). The appearance of induced anisotropy in a planar film is associated with a local mechanical effect. As a result of scratching the normal component of the magnetization vectors appears, which allows us to reveal the domain structure of planar film by the polar Kerr effect.

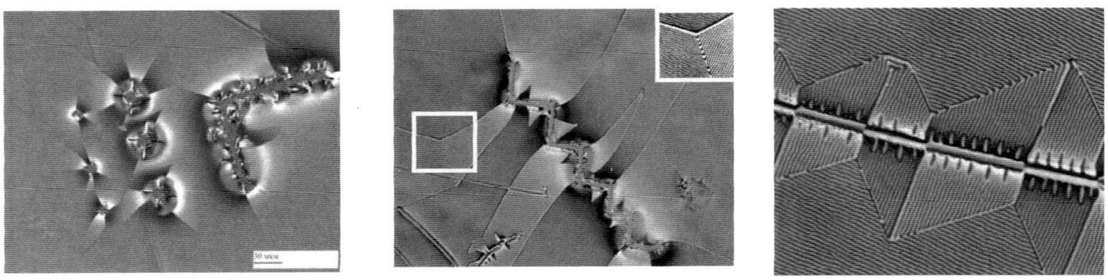

a *b* *c*

Figure 4. MO image of scratch-modified domain structure of planar FG film.

A perpendicular easy axis can be induced by scratching of quadrangular diamond pyramid, loaded up more than 50N on planar FG film. A 180° domain walls, oriented relative to each other at an angle of 90°, were visualized by the polar Kerr effect. The competition between planar film's own cubic and induced uniaxial anisotropy leads to the appearance of a biperiodic domain structure (Figure 4c).

A biperiodic domain structure was described in [9-10]: periodic domain structures appeared in magnetic thin films under the influence of an internal magnetic field directed normally to the film plane and can be presented as a diffraction grating which modulates the linear as well as the nonlinear magneto-optical susceptibilities. In our case, local mechanical scratching led to a biperiodic magnetic structure.

4. Conclusion

The results of our experiments show that defects in magneto-optical bismuth-ferrite – garnet films leads to the formation of closure domains in Bi:FGF with different types of magnetic anisotropy.

The behaviour of the domain structure in the vicinity of a defect in the case of magnetization reversal of uniaxial films is considered. It is shown that after the magnetization reversal of the film, a continuous defect is included in the domain of one sign. Our research demonstrated that the domain structure is sensitive to indentation and scratch tests. It has been shown that local mechanical stresses can lead to the appearance of a biperiodic domain structure not observed earlier in these magnetic materials. The observed effects should be taken into account in magnetoelectronic technology.

Acknowledgement

The authors are grateful to Prof. R.Grechishkin for his valuable assistance in this study.

References

[1] Hubert A and Schafer R 1998 *Magnetic Domains. The Analysis of Magnetic Microstructures* (Springer-Verlag Berlin Heidelberg) 669
[2] FratelloV J, Mnushkina I, Licht S J, and Abbott R R 2005 Growth and characterization of magnetooptic garnet films with planar uniaxial anisotropy *Magneto-Optical Materials for Photonics and Recording* **834** J6.2.1–12
[3] Syvorotka I M, Ubizskii S B, Kucera M, Kuhn M and V′ertesy Z 2001 Growth and characterization of Bi, Pr– and Bi, Sc-substituted lutetium iron garnet films with planar magnetization for magneto-optic visualization *Journal of Physics D: Applied Physics* **34** 1178–1187
[4] Adachi N, Yamaguchi T, Okuda T, Machi T and Koshizuka N 2004 In-plane magnetic anisotropy of (111) and (100) garnet film prepared for magneto-optical indicator *Journal of Magnetism and Magnetic Materials* **272–276** 2255–56
[5] Grechishkin R, Chigirinski S, Gusev M, Cugat O and Dempsey N 2008 Magnetic nanostructures in modern technology. *Magnetic imaging films* (Springer-Berlin) 195–224
[6] Sort J, Concustell A, Menéndez E, Suriñach S, Rao K V, Deevi S C, Baró M D and Nogués J 2006 Periodic arrays of micrometer and sub-micrometer magnetic structures prepared by nanoindentation of a nonmagnetic intermetallic compound *Advanced Materials* **18** 1717–20
[7] Sort J, Bonavina L F, Varea A, Souza C, Botta W J, Kiminami C S, Bolfarini C, Suriñach S, Baró M D and Nogués J 2010 Out-of-plane magnetic patterning based on indentation-induced nanocrystallization of a metallic glass *Small* **6** No. 14, 1543-9
[8] Broese van Groenou 1978 Scratch-modified magnetic domain patterns in garnet films *Applied Physics* **15** 385-92
[9] Dadoenkova N N, Lyubchanskii I L, Lyubchanskii M I and Rasing Th 1999 Nonlinear magneto-optical diffraction from periodic domain structures in magnetic films *Applied Physics Letters* **74** 1880-2
[10] Arzamastseva G V, Lisovskii F V and Mansvetova E G 1998 Polymorphism of biperiodic stripe domain structures *Journal of Experimental and Theoretical Physics Letters* **67** 738-43

IOP Conf. Series: Materials Science and Engineering **503** (2019) 012015 doi:10.1088/1757-899X/503/1/012015

Determination of the electron effective mass for *n*-type germanium by the low-frequency impedance dispersion

O Malyshkina, E Barabanova, K Boitsova, V Klyuev and I Kaplunov

Tver State University, Russia

E-mail: Olga.Malyshkina@mail.ru e

Abstract. We have proposed and approved a method for calculating the electron effective mass in a semiconductor based on the complex impedance analysis. The electron effective mass for Ge *n*–type (doped by Sb) with a specific resistance of 4 Ω·cm (crystallographic directions: [100], [110], [111] and polycrystalline) and 1.5, 7.5, 20 Ω·cm (crystallographic direction [111]) are measured. It is shown that the effective mass depends on the impurity concentration and the crystallographic direction.

1. Introduction

Crystalline germanium (mono- and polycrystals) is one of the classical materials for use in optics. The application of germanium in photonics is about 25 – 30% of the total consumption of this material [1]. Semitransparent mirrors working both without interference coatings, and with them, high-precision Fabry-Perot etalons, acousto-optical elements and other details made from germanium [2–5]. Also, narrow-band interference filters are usually made on Ge substrates.

Impurity germanium crystals are actively used for manufacturing photodetectors recording infrared radiation (IR) [5]. Their work is based on the excitation by a radiation quantum of the charge carrier at the impurity level, passing into the carrierband. For different IR ranges, the appropriate impurity is used. The electrophysical and optical parameters of germanium crystals are correlated. A considerable amount of research of germanium, including optical properties, is carried out by electrophysical methods, which reached considerable perfection in semiconductor electronics. The interrelation between the structure and electrical conductivity (and indirectly, the absorption of radiation) is established not only for single crystals but also for optical polycrystals and for highly pure germanium. The specific electric conductivity of polycrystalline germanium decreases with decreasing crystallite size, which is associated with a decrease in the mobility of charge carriers caused by their scattering at crystallite boundaries. The electrical conductivity in high-purity germanium increases with the decrease in the size of the crystallites, and the lifetime of nonequilibrium charge carriers in it decreases as a result of increase in the concentration of surface electronic states [6, 7].

The newest trend in modern photonics is the creation of new methods of modulation of IR radiation by Ge–based devices, such as modulation by introducing moderate levels of excess carriers in Ge by exciting a diode laser [8]; modulation in waveguide Ge structures on Si substrates by controlling the absorption of free carriers [9, 10]. Ge is a classical semiconductor with a band gap of 0.67 eV [5], hence the effective mass of charge carriers plays an important role in the characterization of its electrophysical properties. At present, the main method for determining the effective mass of charge carriers in a semiconductor is the cyclotron resonance method. At the same time, it characterizes the behavior of free carriers in the field of magnetic forces, whereas in most applications the electric

IOP Conf. Series: Materials Science and Engineering **503** (2019) 012015 doi:10.1088/1757-899X/503/1/012015

forces take place. From such point of view, it is of interest to determine the effective mass of free carriers in AC electric field.

2. Research materials and methods of the experiment

We determined the electron effective mass for samples of *n*–type germanium crystals doped with Sb. Germanium crystals with a diameter of 50–60 mm were grown from the melt by the Czochralski method at Tver State University. Samples of a given crystallographic direction and specific electric resistance were cut from single crystals. Samples under study had the size of 10 × 10 × 5 mm. Charge carrier concentration was determined indirectly using a four-probe method for measuring the electrical resistivity of semiconductors. We used polycrystalline zone-melting purified germanium as a raw material. Doping was carried out with a donor impurity (antimony). With this in view, the doping impurity concentration and the electron concentration are the same.

The corresponding values of the specific resistance (ρ) and the corresponding concentrations of antimony impurities (*n*), as well as the crystallographic directions of the investigated samples of germanium crystals, are shown in Table 1.

Table 1. Characteristics of the germanium samples

Specific resistance, (Ω·cm)	Sb concentrations, (cm^{-3})	Crystallographic direction
4	$3.7\cdot10^{14}$	[100]
4	$3.7\cdot10^{14}$	[110]
4	$3.7\cdot10^{14}$	[111]
4	$3.7\cdot10^{14}$	polycrystalline
1.5	$1\cdot10^{15}$	[111]
7.5	$1{,}7\ 10^{14}$	[111]
20	$8\cdot10^{13}$	[111]

The complex electrical resistance (impedance) of Ge crystals was measured by the LCR method using the Frequency Response Analyzers PSM1735 with Impedance Analysis Interface (Newtons4th Ltd). As electrodes we used two bronze plates between which germanium sample was placed.

The electron effective mass was determined on the basis of analysis of the graphical dependence Z(ω) in the coordinates Z′ and Z″ (the Nyquist coordinates), called the Nyquist plot, or the impedance hodograph [11].

3. Results

During the experiments, it was found that the frequency dependences of the impedance strongly depend on the preliminary surface treatment. A comparison of the impedance hodographs with the literature data [11] showed that the measurements of the complex impedance of the polished sample are more correct. Therefore, the measurements used to calculate the dynamic conductivity were carried out for polished samples. As an example, Figure 1 shows the impedance hodograph for Ge sample with a Sb concentrations $3.7\cdot10^{14}$ cm^{-3} (crystallographic direction [100]). The dotted line on a graph shown on a logarithmic scale (Figure 1b) corresponds to the case Z’ = Z”. It should be noted that the Nyquist plots location below this line for all samples under study. According to [12] it means that samples have a universal capacitance and high losses. The impedance hodograph in the normal scale (Figure 1a) has a "classic" semi-circle shape. This allowed us to use it to determine the effective electron mass.

The effective mass was found from the formula for DC conductivity [12]:

$$\sigma_o = \frac{e^2 n \tau}{m^*}, \tag{1}$$

were e – electron charge, n – concentration of charge carriers (in our case – electrons), m^* – electron effective mass, $\tau = 1/\omega_p$ – relaxation time, $\sigma_o = 1/\rho$.

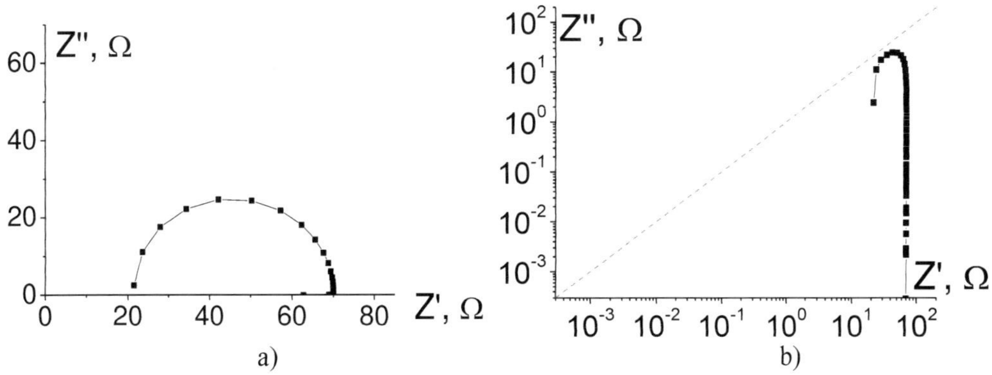

Figure 1. Hodographs in linear (a) and logarithmic (b) scales for n–type Ge sample with a Sb concentrations $3.7\cdot10^{14}$ cm^{-3}. The measurement was carried out in the crystallographic direction [100]

From formula (1), we obtain the expression for electron effective mass:

$$m^* = \frac{e^2 n\tau}{\sigma_o} . \tag{2}$$

To calculate the relaxation time, the frequency ($\omega_p = 2\pi f$) is determined from the impedance hodograph (Figure1a), as the frequency of the point corresponding to the maximum value of the imaginary component of the impedance [12, 13]. For the calculations, the specific resistance (ρ) and the impurity concentration (n) presented in Table 1 were used.

The ratio of the effective electron mass for the germanium samples to the electron mass calculated by formula (2) is shown in Tables 2 and 3. In parentheses the measurement error is given.

Table 2. Calculated values of the ratio of the electron effective mass (m^*) to the electron mass (m_e) for n–type germanium samples with a Sb concentrations $3.7\cdot10^{14}$ cm^{-3}

Crystallographic direction	relaxation frequency (Hz)	m^*/m_e
100	$2.9\cdot10^6$	2.3 (0.2)
110	$2.9\cdot10^6$	2.3 (0.2)
111	$7.4\cdot10^6$	0.9 (0.2)
polycrystalline	$2.9\cdot10^6$	2.3 (0.2)

Table 3. Calculated values of the ratio of the electron effective mass (m^*) to the electron mass (m_e) for n–type germanium samples for the crystallographic direction [111]

Sb concentrations, (cm^{-3})	relaxation frequency (Hz)	m^*/m_e
$10\cdot10^{14}$	$5.4\cdot10^6$	1.2 (0.2)
$3.7\cdot10^{14}$	$7.4\cdot10^6$	0.9 (0.2)
$2\cdot10^{14}$	$2.9\cdot10^6$	2.3 (0.2)
$0.8\cdot10^{14}$	$2.1\cdot10^6$	3.4 (0.2)

As can be seen (Table 2), for germanium with the same impurity concentration, the crystallographic direction [111] differs from the others. Along this direction, the effective mass of electrons is smaller. This means that the periodic potential of the crystal lattice along the direction of the charge carriers (electrons) motion corresponding to the crystallographic orientation [111] is significantly different.

This result correlates with the result obtained for the electron effective mass by the cyclotron resonance method, according to which the electron effective masses in germanium in orthogonally directions are greatly different (3.5 and 0.1 m_e) [14].

4. Conclusion

Investigations of the electron effective mass in *n*–type germanium crystals have shown that it depends both on the crystallographic direction in the crystal and on the impurity concentration. For one crystallographic direction, an increase in the impurity concentration leads to a decrease in the effective mass of the electron. Its value approaches the mass of the free electron. Since the increase in the concentration of the donor impurity leads to an increase in the number of electrons in the conduction band, and they occupy higher energy levels, so their behavior in AC electric field approaches the behavior of free electrons.

The difference of the electron effective mass in the crystallographic direction [111] from other directions most probably is explained by the largest distance between germanium atoms in a given direction (relative to the other directions under study).

References

[1] Anoshin K E, Gasanov A A and Naumov A V 2016 *Non-ferrous metallurgy* **2** 67 (in Russian)
[2] Kaplunov I A, Smirnov Yu M and Kolesnikov A I 2005 *J. Optical Technol.* **72** 214
[3] Kaplunov I A, Kolesnikov A I, and Shaiovich S L 2005 *Crystallografy Reports* **50** 546
[4] Kaplunov I A, Shelopaev A V and Kolesnikov A I 2010 *J. Surface Investigation* **4** 994
[5] Claeys L and Simoen E 2007 Germanium–based technologies: from materials to devices. (Berlin: Elsevier)
[6] Shimanskii A F, Podkopaev O I and Vahrin V V 2011 *J. Siberian Federal University Engineering & Technologies* **4** 542.
[7] Podkopaev O I, Shimanskii A F, Molotkovskaya N O and Kulakovskaya T V 2013 *Phys. Solid State* **55** 872.
[8] Fairley P D and Rutt H N 2000 *J. Phys. D: Appl. Phys.* **33** 2837.
[9] Shen Li, Healy N, Mitchell C J, Penades J S, Nedeljkovic M, Mashanovich G Z and Peacock A C 2015 *Optics Letters* **40** 268.
[10] Soref R, Hendrickson J R and Sweet J 2016 *Optics Express* **24** 9369.
[11] Jonscher A.K 1983 Dielectric relaxation in solids (London: Chelsea Dielectric Press)
[12] Grosse P 1979 Freie Elektronen in Festkorpern (Berlin, Heidelberg, New York: Springer-Verlag)
[13] Jonscher A K 1996 Universal relaxation law (London: Chelsea Dielectric Press).
[14] Fen G I 1958 *Uspekhi Fizicheskikh Nauk* **64** 733.

IOP Conf. Series: Materials Science and Engineering **503** (2019) 012016 doi:10.1088/1757-899X/503/1/012016

Optical properties of zinc-iridium oxide thin films

V Skvortsova, M Zubkins, R Kalendarev, H Arslan and J Purans

Institute of Solid State Physics, University of Latvia, Riga, Latvia

e-mail: vera@cfi.lu.lv

Abstract. We present the results of an investigation of ultraviolet, visible, near-infrared (UV-Vis-NIR) and X-ray absorption spectroscopy absorption spectra for zinc-iridium oxide (Zn-Ir-O) thin films with various iridium concentrations deposited by reactive DC magnetron sputtering. It is found that the absorption spectra of zinc-iridium oxide thin films contain a broad band with maxima at 446 nm and 710 nm in the visible region, bands with maxima at about 1100 nm and 3300 nm, and a low-intensity absorption band at 1570 nm in the near-infrared region. The obtained absorption bands are associated with iridium ions at valences of Ir^{3+}, Ir^{4+} and Ir^{5+}. Changes in the oxidation state of iridium ions from 5+ to 4+ are observed.

1. Introduction

The physics of iridium-based transition metal oxides has sparked significant interest, since iridium oxide has many unique properties that make it an ideal material for applications in electrochromism, bio-sensing and water electrolysis. For most optoelectronic devices such as flat panel displays, it is essential to use a transparent electrode consisting of a thin film of a transparent conducting oxide (TCO) semiconductor. Kawazoe et al. [1] reported *p*-type conductivity in highly transparent thin films of copper aluminum oxide. In this publication, a variety of *p*-TCOs were investigated. Despite these results, the insufficient transparency and low conductivity of the *p*-TCOs continue to restrict their technological applications [2]. Several precious-metal-based transparent oxides have also shown *p*-type conductivity, including polycrystalline ZnM_2O_4 (M= Ir, Rh. Co) films [3-5]. An increase in the Ir concentration of the Zn-Ir-O thin films from 12 to 16 at. % changes the conductivity type from *n*- to *p*-type [6].

Iridium has the ability to form compounds in many different oxidation states, the most stable of which are +3 and +4. The mixed-valence nature of Ir^{3+} and Ir^{4+} ions has been observed in several compounds; Ir^{5+} and Ir^{6+} can also be obtained [7-10].

Theoretical calculations of the substitution defects of $Ir^{2+}{}_{Zn}$ and $Ir^{3+}{}_{Zn}$ in Ir-doped ZnO predict localised energy states in the band gap, which is expected to reduce the transmittance of ZnO: Ir films [11, 12]. A decrease in the transmittance of the films is observed experimentally [6, 13].

In this paper, UV-Vis-NIR and X-ray absorption spectroscopy are used to investigate the valence states of Ir ions in Zn-Ir-O films deposited by reactive DC magnetron co-sputtering.

2. Experimental methods and materials

Zinc-iridium oxide thin films were deposited on glass, Si and polyimide tape substrates by reactive DC magnetron co-sputtering in an Ar + O_2 atmosphere. Two types of sputtering methods were used to deposit the films studied here. The detailed experimental procedure of both methods can be found in [6, 14, 15]. An elemental analysis of the films was carried out using an Eagle III X-ray fluorescence spectrometer. Since it is difficult to quantify elements lighter than sodium with XRF, our measurements include only the Zn and Ir concentration ratios. To determine the influence of the substrate temperature on the structure and properties of the films, two sets of samples were deposited: one set without intentional substrate heating during the deposition process, and a second with additional heating at a temperature of 300°C.

The Zn K-edge (9659 eV) and Ir L_3-edge (11215 eV) X-ray absorption spectra (XAS) were measured in transmission mode at the SOLEIL synchrotron bending-magnet SAMBA beamline [16] under ambient conditions. More details of this experiment can be found in [6].

For the UV-Vis absorption spectra measurements at room temperature, we employed a Specord 210" (Analytikjena), a double-beam spectrophotometer operating in the spectral region of 190–1100 nm. The FTIR spectra were also measured at RT using a VERTEX 80v vacuum FTIR spectrometer with an actively aligned UltraScan™ interferometer

3. Results and discussion

The absorption spectra for zinc-iridium oxide thin films deposited on a glass substrate at various iridium concentrations are shown in Figure 1. Zinc-iridium oxide thin films are transparent in the visible region. An increase in the iridium concentration leads to the attenuation of film transparency. This increase also shifts the absorption edge towards the shorter wavelengths compared to crystalline ZnO, leading to higher absorption coefficient in the visible range. The results of curve fitting (the band shapes are assumed to be Gaussian) give a bands with maxima at 446 and 710 nm. The intensities of the bands increase with the iridium concentration of the thin films.

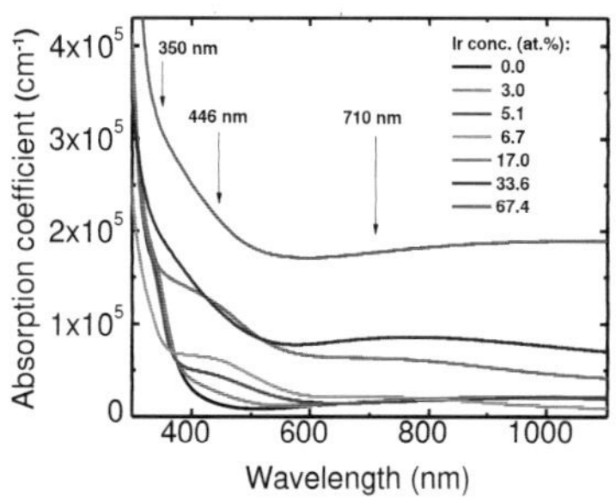

Figure 1. Absorption spectra of Zn-Ir-O thin films with different iridium concentrations.

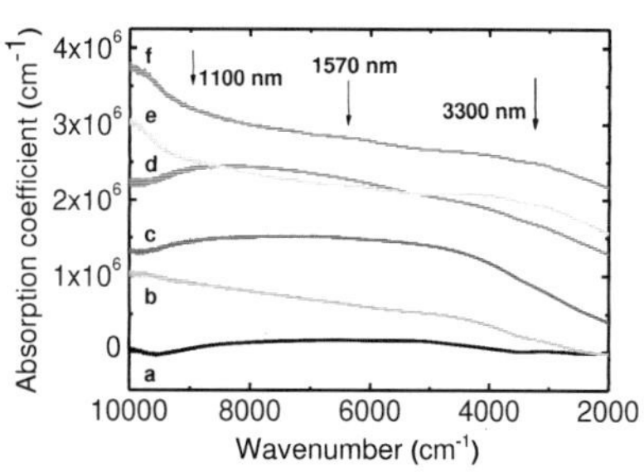

Figure 2. Near infrared absorption spectra of Zn-Ir-O thin films with different iridium concentrations (at.%): 7 (a), 16 (b), 29 (c), 41(d), 47 (e), and 54 (f).

In the near-infrared and infrared regions, the absorption spectra of zinc-iridium oxide thin films with different iridium ion concentrations show two broad bands with maxima at about 1100 and 3330 nm (Figure 2). The intensity of the 1100 nm band in the zinc-iridium oxide thin films increases with the iridium concentration, reaching a maximum at an iridium concentration of 29%. A further increase in the iridium concentration to 41% leads to an increase in intensity of the 3330 nm band. At an iridium concentration of between 47% and 56%, the ratio of the intensities of the 1100 and 3330 nm bands remains unchanged. A low-intensity absorption band at ~1570 nm is also observed.

The 5d transition-metal ions are known to be present in different valences in many compounds. A very complex situation can occur when a certain ion is present in more than one valence state and when these ions occupy both tetrahedral and octahedral sites. The valence and site occupation of these ions are mainly determined by the growth conditions and the current impurities

It has been reported in the literature that the low-spin state of Ir 5d electrons is preferred. Comparing the obtained results with the Tanabe-Sugano diagram [17], we conclude that the absorption spectra bands correspond to the Ir^{3+} ion transition from the ground to the excitation state: band 3300 nm $^1A_{1g}\rightarrow{}^5T_2$ transition, band 1100 nm $^1A_{1g}\rightarrow{}^3T_{1g}$ transition, band 710 nm $^1A_{1g}\rightarrow{}^3T_{2g}$ transition, and band 446 nm $^1A_{1g}\rightarrow{}^1T_{1g}$ transition. The transitions $^1A_{1g}\rightarrow{}^5T_{2g}$, $^1A_{1g}\rightarrow{}^3T_{1g}$ and $^1A_{1g}\rightarrow{}^3T_{2g}$ are forbidden and have low intensities. The restrictions imposed by the selection rule are removed due to the exchange interaction. The transition $^1A_{1g} \rightarrow {}^1T_{1g}$ is allowed. An analogous absorption band associated with the trivalent ion iridium has observed in prior papers [18].

Tetravalent iridium has the electronic configuration of $5d^5$, which corresponds to a half-filled d-sub-shell and is particularly most stable. In the ground state of the d^5 ion in the low-spin state, ^6S transforms into the $^2T_{2g}$ state. The transitions from the ground to the quartet and sextet states are forbidden, and hence are too weak. The bands observed in the absorption spectra of zinc-iridium oxide thin films may belong to the transitions from the ground state $^2T_{2g} \rightarrow {}^6A_{1g}$ (band at 3330 nm), $^2T_{2g} \rightarrow {}^4T_{1g}$ (band at 710 nm) and $^2T_{2g} \rightarrow {}^4T_{2g}$ (band at 446 nm) to the excited state of the iridium ions. The intensity of the 710 nm band is less than that of the 446 nm band. A significant intensity increase in the band at 710 nm is observed at an iridium concentration of more than 20%. This indicates the oxidation of Ir^{3+} to Ir^{4+}. The same process was observed in [19].

Pentavalent iridium has the electronic configuration d^4. The ground state of the d^4 ion in the low-spin state is $^3T_1$. The transitions from the ground to the excited ^5E, $^1T_2$, ^1E and $^1A_1$ states are forbidden, and hence are too weak. The allowed transitions ^3E, $^3T_1$, $^3T_2$, $^3A_1$ and $^3A_2$ lie in the region above ~330 nm (~30000 cm^{-1}) [20]. A low-intensity absorption band is observed for thin films of zinc-iridium oxide in the near-infrared region at a maximum of ~ 1570 nm. The d^4 electron configuration is shown by Cr^{2+} ions, and in the absorption spectra for the $ZnS:Cr^{2+}$ compound, its observed band is at ~1600 nm [21]. We assume that the band at 1570 nm in the zinc-iridium oxide films is connected with the $^3T_{1g} \rightarrow {}^5E$ transition of the Ir^{5+} ions.

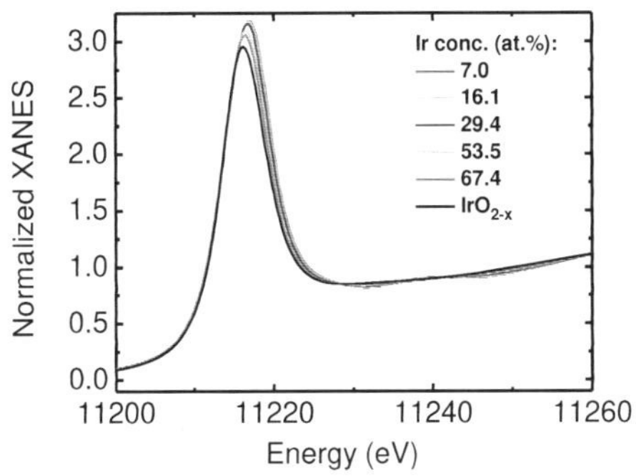

Figure 3. XANES spectra at the Ir L_3-edge of the deposited films.

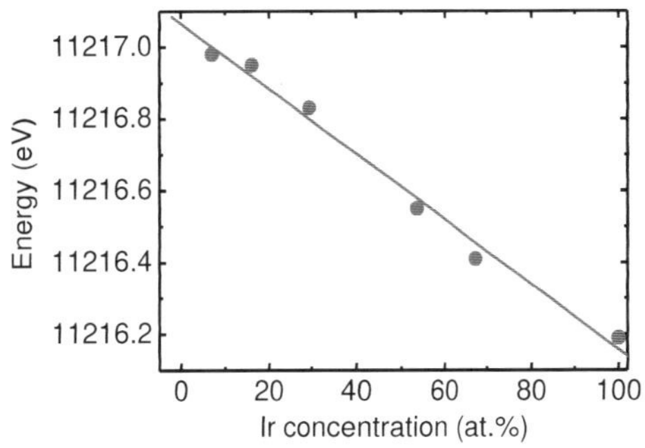

Figure 4. Plot of the maximum position energy of the XANES spectra as a function of iridium concentration in the zinc-iridium oxide films

X-ray absorption spectroscopy (XAS) is typically one of the most optimal techniques for this work. In order to investigate the local atomic structure, we need the bounding characteristics, oxidation and valance state of the atom of interest. This technique can be separated into two regimes based on the absorbed energy: (i) the X-ray absorption near the near-edge structure (XANES), which extends up to nearly 40 eV and is very sensitive to the charge density (i.e. the evolution of the oxidation sate) and bounding traits of the absorbing atom; and (ii) the extended X-ray absorption fine structure (EXAFS), which is specific to the local environment of the selected atom for energies higher than 40 eV. As emphasised above, in this research we investigate the oxidation state of the Ir atom by analysing the bounding energy shift of core level electrons [22]. For the XANES L-edge ($L_{2,3}$, $2P_{1/2}$ and $2P_{3/2}$ states) absorption profiles, the magnitude of the absorption intensity indicates the available vacant states in the d-orbital, and it is therefore proportional to the oxidation state of the absorbing atoms [23]. Figure 3 depicts the XANES spectra for the Ir L_3-edge from the zinc-iridium oxide thin film. A plot of the energy position as a function of iridium concentration in zinc-iridium oxide thin films is given in Figure 4. The distribution of energy positions is significantly larger in the higher oxidation states. We observe a change in the oxidation state of iridium ions in zinc-iridium oxide thin films from 5+ to 4+, as indicated in [24].

4. Conclusion

The research has been applied to the ultraviolet, visible and near infrared (UV-Vis-NIR) absorption spectra of zinc-iridium oxide (Zn-Ir-O) thin films with various iridium concentrations deposited by reactive DC magnetron sputtering It was found out that the absorption spectra of zinc-iridium oxide thin films include a broad band with maxima 446 and 710 nm in the visible region and bands with maxima at about 1100, 3300 nm and the low intensity absorption band 1570 nm in near infrared region. The obtained bands are shown to be associated with iridium ions with different valence states Ir^{3+}, Ir^{4+} and Ir^{5+}. Changes in the oxidation state of iridium ions from 5+ to 4+ are observed.

Acknowledgment

This work was performed with support from VMTKC Project 18, under agreement No. 1.2.1.1/16/A/005.

References

[1] Kawazoe H, Yasukawa M, Hyodo H, Kurita M, Yanagi H and Hosono H 1997 *Nature* **389**(6654) 939
[2] Klein A 2013 *J. Am. Ceram. Soc.* **96** 331
[3] Ohta H, Mizoguchi H, Hirano M, Narushima S, Kamiya T and Hasono H 2003 *Applied Physics Letters* **82** 823
[4] Dekkers M, Rijnders G, and Blank D H A 2007 *Applied Physics Letters* **90** 021903 1-3
[5] Kim H J, Song I, Sim C H, Kim D, Kim Y, Ihm E, Choo W 2004 *J. of Applied Physic* **95** 7387
[6] Zubkins M, Kalendarev R, Gabrusenoks J, Plaude A, Zitolo A, Anspoks A, Pudzs K, K. Vilnis, Azens A, Purans J 2017 *Thin Solid Films* **636** 694
[7] Radaelli P G, Horibe Y, Gutmann M J, Ishibashi H, Chen C H, Ibberson R M, Koyama Y, Hor J S, Kiryukhin V and Cheong S-W 2002 *Nature* **416** 155
[8] Kumagai K, Kakuyanagi R, Endoh R and Nagata S 2000 *Physica* **C 341** 741
[9] Frenzen S and Müller-Buschbaum Hk 1996 *Z. Naturforsch. B: J. Chem. Sci.* **51** 822
[10] Mugavero S J, Smith M D, Yoon W-S and zur Loye H-S 2009 *Angew. Chem. Int. Ed.* **48** 215
[11] Muñoz Ramo D and Bristowe P D 2014 *Thin Solid Films* **555** 112
[12] Muñoz Ramo D and Bristow P D 2014 *J. Chemical Physics* **141**, 084704
[13] Michail G, Kambylafka V, Kortidis I, Tsagaraki K, Androulidaki M, Kiriakidis G, Binas V, Modreanu M 2016 *Thin Solid Films* **617** (Part A) 3
[14] Zubkins M, Kalendarev R, Gabrusenoks J, Vilnis K, Azens A and Purans 2014 *Phys. Status Solidi C* **11**(9) 1493
[15] Zubkins M, Kalendarev R, Gabrusenoks J, Smits K, Kundzins K, Vilnis K, Azens A and Purans J 2015 *IOP Conf. Series: Materials Science and Engineering* **77** 1012035
[16] Briois V, Fonda E, Belin S, Barthe L, La Fontaine C, Langlois F, Ribbens M and Villain F 2011, *EDP Sciences* 41
[17] Tanabe Y and Sugano S 1959 *J. Phys. Soc. Japan* **9** 766
[18] Grochowski J, Kaminska E, Piotrowska A, Dynowska E, Dluzewski P, Dyczewski J, Szczepanska A and Kazmierczak P 2012 *Phys. Status Solidi C* **9** 1504
[19] Castillo-Blum S E, Richens D T and Sykes A G 1989 *Inorganic Chemistry* **28** 954
[20] Anghel S, Brik M G, Mac C-G, Sushkevich K and Kulyuk L 2015 *J. Luminescence* **166** 282
[21] Karhu E A, Ildstad C R., Poggio S, Furtula V, Tolstik N, Sorokina I T, Belbruno J and Gibson U J 2016 *Optical Materials Express* **9** 2947
[22] Hillman A R, Skopeka M A and Gurman S J 2011 *Phys. Chem. Chem. Phys.* **13** 5252
[23] Sinfelt J H and Meitzner G D 1993 *Acc.Chem.Res.* **26** (1) 1
[24] Choy J H, Kim D-K, Hwang S-H, Demazeau D and Jung D-Y 1995 *J. Am. Chem. Soc.* **117** 8557

IOP Conf. Series: Materials Science and Engineering **503** (2019) 012017 doi:10.1088/1757-899X/503/1/012017

Electronic processes in doped ZnO nanopowders

Larisa Grigorjeva, Aleksejs Zolotarjovs, Ivita Bite, Jurgis Grube, Donats Millers

Institute of Solid State Physics UL, Kengaraga, 8, LV-1063, Riga, Latvia

lgrig@latnet.lv

Abstract. ZnO nanocrystals, undoped and doped with Iridium or Indium were prepared by solar irradiation in Heliotron reactor in PROMES CNRS facilities, France. The comparative analysis of the excitonic spectra of ZnO single crystal and ZnO nanocrystals (NCs) doped with In and Ir was performed. It is shown that the excitonic processes in Ir doped nanocrystals coincide well with electronic processes in undoped NC and single crystal; however, the electronic processes in In-doped nanocrystals are significantly different from those in single crystal. The radioluminescence spectra of ZnO:In was analysed and additional luminescence band at ~3.18 eV was detected due to In-doping. The luminescence decay time depends on In concentration in nanocrystals and is significantly less in ZnO:In compared with undoped nanocrystals. The fast scintillation of ZnO:In makes this material promising for application

1. Introduction

ZnO has excellent luminescence characteristics - especially fast intrinsic luminescence [1] and application for fast scintillators [2]. The electronic processes in ZnO crystal, nanocrystals (NCs), coatings and ceramics are vastly different and depend on various factors: doping concentration, intrinsic defects, electron/hole traps distribution and others. Donor-acceptor pair (DAP) luminescence can also serve as a direct probe of the acceptors in ZnO [3]. The luminescence spectroscopy is classically used as an informative method for the electronic processes and doping effect studies in ZnO.

Another advantage of this material is that there are various already established methods of synthesis of nanostructured ZnO with many more under development. In the present study, the relatively new synthesis method is used: solar physical vapour deposition (SPVD) [4]. Apart from NCs, for the practical applications, the ZnO ceramics are also promising. It was proven that semi-transparent ZnO ceramics can be obtained from undoped and In, Ga doped powders [5,6] and the excitonic luminescence of such materials is in the sub-nanosecond time region and therefore is perspective for use in fast scintillators. In the present study, we evaluate the possibilities of low temperature excitonic luminescence in ZnO, ZnO:In and ZnO:Ir NCs. The study is based on measurements of spectra and decay kinetics at 300K of ZnO and ZnO:In NCs with the idea of using of such NCs as raw material for the production of fast scintillators.

2. Experimental

2.1. Samples preparation

ZnO NCs undoped and doped with Ir, In were prepared by evaporation using condensed solar light in Heliotron reactor in Font Romeu, PROMES CNRS, France [2]. The concentration of dopants was changed by introducing of various concentrations (wt.%) of oxides of the corresponding dopants in the target. In concentration was 0.5 wt.%, 1.0 wt.%, 3.0 wt.%. Ir concentration used in this study was 3.0 wt.%.

2.2. Experiment

Radioluminescence was measured under x-ray excitation (30 kV; 10 mA, W-anode). Photoluminescence was excited with 266 nm; 1 ns laser pulses. The luminescence spectra and decay kinetics were measured with Streak-scope C4334 (HAMAMATSU) under 30 ps pulses with 330 nm excitation. The spectra registration time gates were 1 ns and 5 ns. The pellets with powders were attached to a metallic surface placed inside a closed cycle helium gas cryostat (Advanced Research Systems, Model CSW-202). The temperature of measurements was in the region of 10 K-300 K. The radioluminescence spectra were recorded using the Andor Shamrock B-303i spectrograph with a CCD camera (Andor DU-401A-BV).

3. Results

3.1. Low temperature radioluminescence

The low-temperature luminescence spectra of ZnO are studied in detail and the interpretation of the main bands is fairly well known [3]. The low-temperature exciton spectra already can be used to get information about the quality of the material. However, the search for hole centers in ZnO is still undergoing and the presence of DAP luminescence shows the formation of hole centers. In the present paper we compare the excitonic spectra of the ZnO single crystal and ZnO NCs doped with In and Ir (Fig.1). The main band in the single crystal and ZnO:Ir is located at 3.358 eV is ExD^0, a band at 3.33 eV is due to 1LOFEx and these bands are well resolved. In addition, the band at ~3.22 eV in ZnO:Ir may be caused by donor-acceptor pairs (DAP). Thermal quenching of DAP luminescence is in T<100 K [2]. In the case of ZnO:In, the radioluminescence spectrum has a distinctive nature. Excitonic bands of ExD^0 and 1LOFEx in ZnO:In are hidden under the edge of fundamental absorption and/or under a wide band associated with the In - doping.

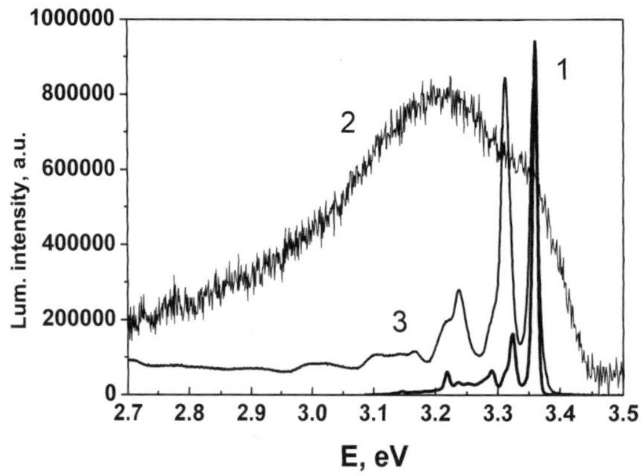

Figure 1. Radioluminescence spectra at 9 K of ZnO single crystal (1), In doped (3 wt.%) (2) and Ir (3 wt.%) (3) doped NC.

3.2. Fast luminescence of ZnO and ZnO:In NC at 300K

The fast luminescence spectra and decay kinetics at 300 K were measured under 266 nm excitation in 1 ns time gate. The spectra are shown in Fig.2. Analysis of spectrum gives the basis to conclude that the spectra of ZnO and ZnO:In differs significantly and therefore the decomposition of the spectra into Gaussian components has been carried out. The results are presented in Fig.3, a,b,c. One band peaking at 3.2 eV with FWHM $\approx$ 0.12 eV is observed in undoped NCs. The high-energy edge of the luminescence band may be slightly distorted due to the reabsorption associated with the edge of fundamental absorption.

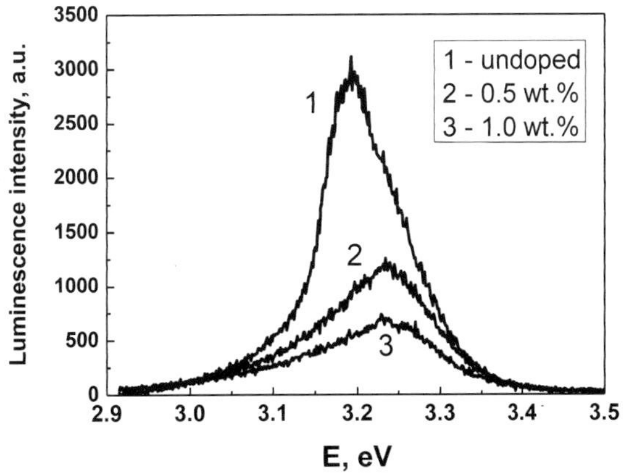

Figure 2. Luminescence spectra of ZnO and ZnO:In at 300 K in 1 ns time range and 266 nm excitation

On the other hand, in ZnO:In doped NCs the spectra consists of two Gaussian components. The additional component position is shifted to low-energy region and the FWHM is increased when compared to undoped NCs. The additional peak is located at 3.18 eV and FWHM is 0.24 eV. The obtained results show that the additional fast luminescence peak in In - doped NCs can be used to explain the effect of scintillation peak broadening in doped samples.

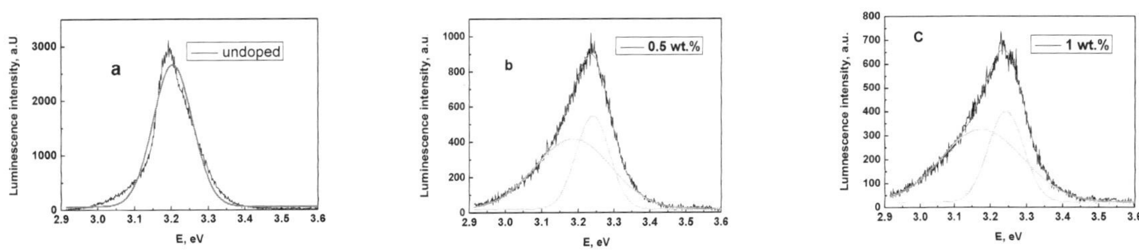

Figure 3. The results of decomposition of luminescence band into Gaussian components for undoped (a) and In-doped (b and c).

3.3 Luminescence decay kinetics in ZnO:In NC

The ZnO:In is considered as a promising material for fast scintillators, so the scintillation decay time estimation is important. In Fig.4 the decay kinetics for undoped NCs and ZnO:In with 0.5 and 1.0 wt.% concentration are shown. The decay process is not exponential, and this can be explained by the two luminescence centers (Fig.4). Doping concentration affects the decay time which decreases with an increase of In-concentration.

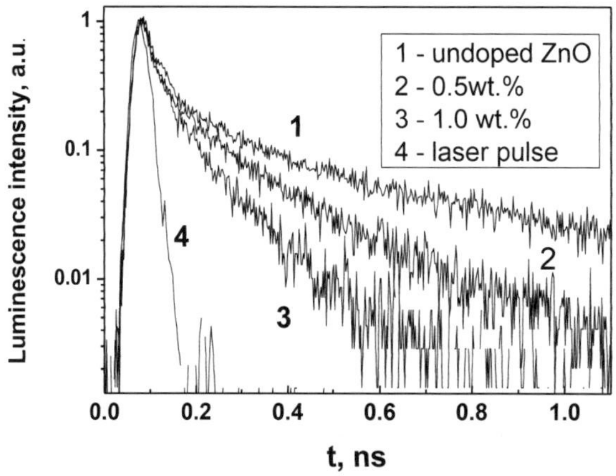

Figure 4. Luminescence decay kinetics of ZnO and ZnO:In NC in spectral region 3.24 eV.

4. Summary

The electronic processes in In doped nanocrystals are significantly different from those in the single crystal. Excitonic bands of ExD^0 and 1LOFEx in ZnO:In are hidden under the edge of fundamental absorption and under a wide band associated with the In-doping. The additional fast luminescence peak was obtained in In-doped NC and explains the effect of scintillation peak broadening in doped samples. The luminescence decay time depends on In concentration in NC and is significantly less in ZnO:In when compared with undoped NC

Acknowledgments

The financial support of research grants ERA.NET RUS_ST20170-051 and SFERA II project "Transnational Access activities" (EU 7th Framework Programme Grant Agreement N312643). This work was partly supported by the Russian Foundation for Basic Research, project no. 18-52-76002.

References

[1] Özgür Ü, Alivov Ya, Liu C, Teke A, Reshchikov M A, Doğan S, Avrutin V,. Cho S-J and Morkoç H 2005 *J. Appl. Phys.* **98**, 041301.
[2] Wilkinson J, Ucer K B, Williams R.T. 2004 *Radiation Measurements* **38,** 501.
[3] Meyer B K, Alves H, Hofmann D M, Kriegseis W, Forster D, Bertram F, Christen J, Hoffmann A, Strasburg M, Dworzak M, Haboeck U and Rodina A V 2004 *Phys. Stat. Sol. (b)* **241**, No. 2, 231.
[4] Smits K, Grigorjeva L, Millers D, Kundzins K, Ignatans R, Grabis J and Monty C 2014 *Optical Materials*, **37(C),** 251.
[5] Demidenko V A, Gorokhova E T, Kristich O A, Mirhin S B and Rodnyi P A 2007 *Radiation Measurements,* **42**, 549.
[6] Chernenko K, Gorokhova E, Eron'ko S, Sandulenko A, Venevtsev I, Wieczorek H and Rodnyi P 2018 *IEEE Transactions on Nuclear Science*, TNS-00113.

IOP Conf. Series: Materials Science and Engineering **503** (2019) 012018 doi:10.1088/1757-899X/503/1/012018

Pressure-induced structural changes in α-MoO_3 probed by X-ray absorption spectroscopy

Inga Jonane[1], Andris Anspoks[1], Lucie Nataf[2], François Baudelet[2], Tetsuo Irifune[3] and Alexei Kuzmin[1]

[1] Institute of Solid State Physics, University of Latvia, Kengaraga street 8, LV-1063 Riga, Latvia

[2] Synchrotron SOLEIL, l'Orme des Merisiers, Saint-Aubin, BP 48, 91192 Gif-sur-Yvette, France

[3] Geodynamics Research Center, Ehime University, 2-5 Bunkyo-cho, Matsuyama, Ehime 790-8577, Japan

E-mail: inga.jonane@cfi.lu.lv

Abstract. Energy-dispersive X-ray absorption spectroscopy at the Mo K-edge was used to study pressure-induced (up to 36 GPa) changes in the local atomic structure of 2D layered oxide α-MoO_3. A linear combination analysis based on the low and high-pressure X-ray absorption near edge structure (XANES) spectra shows clear evidence of two high-pressure phases, existing at 18-25 GPa and above 32 GPa. The first transition is due to gradual decrease of the interlayer gap, whereas the second one – to its collapse and oxide structure reconstruction. The local atomic structure around molybdenum atoms at 0.2, 18.5 and 35.6 GPa was determined from the extended X-ray absorption fine structure (EXAFS) using reverse Monte Carlo calculations.

1. Introduction

After the discovery of graphene, other members of 2D layered materials family, including molybdenum trioxide (MoO_3), have been intensively investigated due to their unique properties [1]. MoO_3 is known to exist in six different phases: α-MoO_3, h-MoO_3, β-MoO_3, β'-MoO_3, MoO_3-II and MoO_3-III [2, 3]. Only orthorhombic α-phase (space group $Pbnm$) is the thermodynamically stable at ambient conditions. The structure of α-MoO_3 (Fig. 1) is composed of strongly distorted MoO_6 octahedra, which share edges and form chains linked into layers stacked on top of each other with rather weak van der Waals forces [4]. The oxidation state of molybdenum ions is +6, corresponding to the electronic configuration $4d^0$, which is responsable for their displacement from the octahedron centre due to the second-order Jahn-Teller effect [5].

α-MoO_3 attracts much attention because of its electrochromic and photochromic properties [6, 7]. Layered structure makes it prospective for applications in solar cells [8], catalysis [9, 10] and gas sensing devices [11, 12, 13]. At the same time, technologically relevant electric and optical properties can be affected when high pressure is applied.

The high-pressure behaviour of microcrystalline α-MoO_3 has been investigated in the past by angle-dispersive synchrotron X-ray powder diffraction and Raman spectroscopy techniques up to 43 GPa [3]. Two phase transitions were found at room temperature – to monoclinic MoO_3-II phase ($P2_1/m$) at ~12 GPa and to monoclinic MoO_3-III phase ($P2_1/c$) at ~25 GPa (Fig. 1). Note that MoO_3-II phase was observed before only at high-pressure and high-temperature in [14].

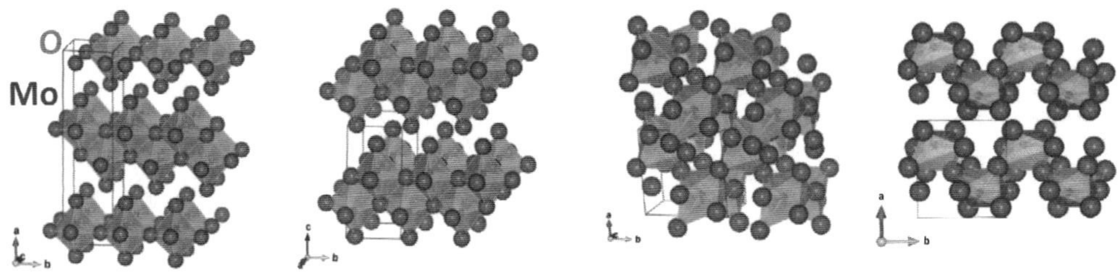

Figure 1. Structures of molybdenum trioxide: orthorhombic α-MoO_3 ($Pbnm$) [4], monoclinic MoO_3-II phase ($P2_1/m$) and monoclinic MoO_3-III phase ($P2_1/c$) [3, 14]. Two ways of structure representation are shown for MoO_3-III using MoO_7 (left) and MoO_6 (right) polyhedra.

While the first two phases, α-MoO_3 and MoO_3-II, are quite similar from the local environment point of view, a collapse of the interlayer gap occurs in MoO_3-III phase [3].

In this study, we report on the pressure-dependent X-ray absorption spectroscopy (XAS) study of α-MoO_3 up to ~36 GPa to evaluate the influence of pressure on the local atomic and electronic structure. Since XAS is extremely sensitive to the distortions in the local environment around the absorbing atom, its spectral features in X-ray absorption near edge structure (XANES) and extended X-ray absorption fine structure (EXAFS) regions can be used to monitor and identify structural changes during compression.

2. Experimental

Pressure-dependent Mo K-edge (20000 eV) X-ray absorption spectroscopy of microcrystalline α-MoO_3 was performed at room temperature using energy-dispersive setup of bending-magnet ODE beamline [15] located at SOLEIL synchrotron facility. The SOLEIL synchrotron operated in the top-up mode with the energy E=2.75 GeV and current I=450 mA. A dispersive polychromator Si(311) was employed to focus X-rays, and the X-ray intensity was measured by a Princeton Instruments PIXIS-400 CCD camera coupled with a scintillator. A membrane-type nano-polycrystalline diamond anvil cell (NDAC) [16] was used to control the sample pressure up to 36 GPa. The silicon oil was used as a pressure transmitting media. The pressure in the cell was determined from a position of the R_1-line of ruby fluorescence excited at 473 nm.

3. Results and discussion

Pressure dependence of the Mo K-edge XANES in Fig. 2(a) indicates the presence of three phases with transition points above 9.9 GPa and 26.3 GPa. After the pressure is released, the system returns back to the α-phase. A phase fraction was evaluated from XANES in the energy range from 19985 to 20075 eV at each pressure using a linear combination analysis (LCA). The experimental XANES spectra at P=0.2, 18.5 and 32.4 GPa were selected as a reference spectra corresponding to pure α-MoO_3, MoO_3-II, MoO_3-III phases, respectively. The obtained results are plotted in Fig. 2(b) and are in good agreement with X-ray diffraction data [3]. In the pressure range between ~10 and 15 GPa, the two phases, α-MoO_3 and MoO_3-II, coexist in a close ratio, whereas MoO_3-III phase appears above ~26 GPa, leading to a change of the XANES.

These findings are also supported by a behavior of the Mo K-edge EXAFS (Fig. 3). The transition from MoO_3-II to MoO_3-III phase is accompanied by a change of the Mo–O–Mo angles between neighbouring molybdenum-oxygen polyhedra from ~144° and ~169° in MoO_3-II [14] to ~149° in MoO_3-III [3]. The absence of the Mo–O–Mo angle equal to 169° is responsible for a

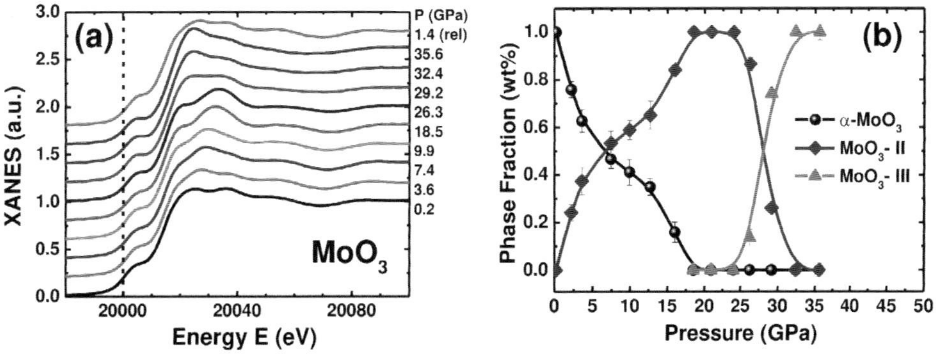

Figure 2. (a) Pressure dependence of the Mo K-edge X-ray absorption near edge structure (XANES) in molybdenum trioxide. (b) Variation of the phase fraction as a function of applied pressure obtained from XANES by the linear combination analysis.

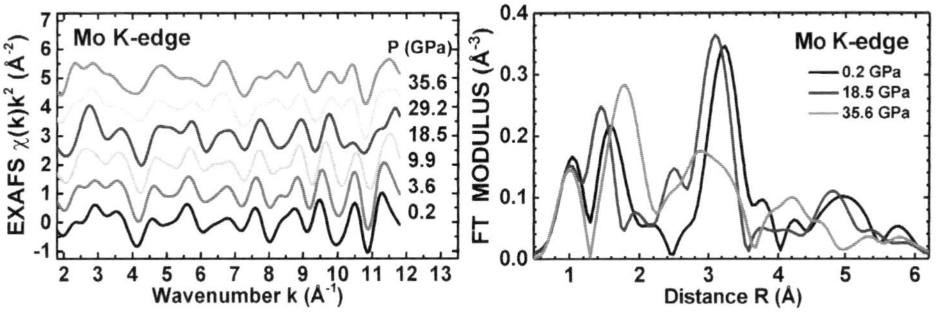

Figure 3. Pressure dependence of the Mo K-edge EXAFS spectra $\chi(k)k^2$ and their Fourier transforms (FTs) in molybdenum trioxide.

decrease of the second shell peak (at ~3.2 Å) amplitude in the Fourier transforms of the EXAFS spectra due to a reduction of the multiple-scattering (MS) effects.

To follow pressure induced changes in details, the analysis of EXAFS spectra beyond the first coordination shell of molybdenum was performed using reverse Monte Carlo (RMC) calculations [17], including the MS effects up to the 6th order. Structural models for α-MoO_3, MoO_3-II and MoO_3-III phases based on the diffraction data from [2, 3] were used as initial atomic configurations and resulted in good agreement with the experimental EXAFS data. The atomic coordinates obtained from RMC simulations were employed to calculate the partial radial distribution functions (RDFs) $g_{\text{Mo}-\text{O}}(R)$ and $g_{\text{Mo}-\text{Mo}}(R)$ for α-MoO_3 (0.2 GPa), MoO_3-II (at 18.5 GPa) and MoO_3-III (35.6 GPa) phases (Fig. 4). Note that the peaks in the RDFs $g_{\text{Mo}-\text{O}}(R)$ up to about 2.6 Å correspond to the nearest oxygen atoms, forming the coordination polyhedra.

At ambient conditions, there are three groups of two oxygen atoms each at ~1.70, 1.96 and 2.26 Å, which form strongly distorted MoO_6 octahedra in α-MoO_3. Already a small pressure of about 0.2 GPa induces lattice contraction, leading to an increase of the MoO_6 octahedra distortion and splitting of some Mo Mo peaks. In MoO_3-II phase at 18.5 GPa, the layered structure persists, MoO_6 octahedra remain distorted and do not differ significantly from that in α-MoO_3, as is expected from diffraction data [2, 3]. At high pressure (35.6 GPa), the collapse of layered structure leads to an increase of molybdenum coordination. Six nearest oxygen atoms from the same layer are responsible for the peaks at ~1.68, 1.88 and 2.12 Å, whereas the 7th oxygen atom bridging two layers is located at ~2.48 Å. The high pressure modifies also the

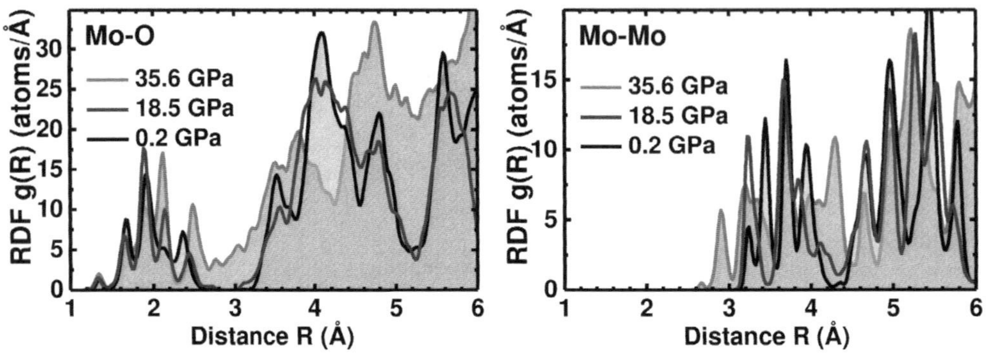

Figure 4. Radial distribution functions for Mo–O and Mo–Mo atom pairs in α-MoO_3 (at 0.2 GPa), MoO_3-II (at 18.5 GPa) and MoO_3-III (at 35.6 GPa), reconstructed by RMC method.

Mo–Mo distribution, leading to a shortening of the distance between neighboring molybdenum-oxygen polyhedra connected by edges (peaks at 3.25 and 2.9 Å).

4. Conclusions

Two phase transitions from α-MoO_3 to MoO_3-II and, next, to MoO_3-III were observed by the Mo K-edge X-ray absorption spectroscopy at room temperature in the pressure range from 0 to 36 GPa. Gradual decrease of the interlayer spacing is responsible for the first transition, whereas a collapse of the interlayer gap occurs when the oxide transforms to MoO_3-III phase. The change of the molybdenum local environment upon phase transitions was determined by RMC method. After the pressure is released, the oxide restores its structure back to the orthorhombic α-phase.

Acknowledgments

The authors are grateful to Prof. Alain Polian for providing NDAC cell. The research leading to this result has been supported by the project CALIPSOplus under the Grant Agreement 730872 from the EU Framework Programme for Research and Innovation HORIZON 2020. The work was supported by philanthropist MikroTik and administrated by the University of Latvia Foundation.

References

[1] Geim A K and Grigorieva I V 2013 *Nature* **499** 419–425
[2] Parise J, McCarron III E, Von Dreele R and Goldstone J 1991 *J. Solid State Chem.* **93** 193–201
[3] Liu D, Lei W, Hao J, Liu D, Liu B, Wang X, Chen X, Cui Q, Zou G, Liu J *et al.* 2009 *J. Appl. Phys.* **105** 023513
[4] Åsbrink S, Kihlborg L and Malinowski M 1988 *J. Appl. Crystallogr.* **21** 960–962
[5] Kunz M and Brown I D 1995 *J. Solid State Chem.* **115** 395–406
[6] Yao J, Hashimoto K and Fujishima A 1992 *Nature* **355** 624
[7] He T and Yao J 2003 *J. Photochem. Photobiol. C* **4** 125–143
[8] Girotto C, Voroshazi E, Cheyns D, Heremans P and Rand B P 2011 *ACS Appl. Mater. Interfaces* **3** 3244–3247
[9] Labanowska M 2001 *ChemPhysChem* **2** 712–731
[10] Shakir I, Shahid M and Kang D J 2010 *Chem. Commun.* **46** 4324–4326
[11] Sunu S, Prabhu E, Jayaraman V, Gnanasekar K, Seshagiri T and Gnanasekaran T 2004 *Sens. Actuator B Chem.* **101** 161–174
[12] Comini E, Yubao L, Brando Y and Sberveglieri G 2005 *Chem. Phys. Lett.* **407** 368–371
[13] Ji F, Ren X, Zheng X, Liu Y, Pang L, Jiang J and Liu S F 2016 *Nanoscale* **8** 8696–8703
[14] McCarron III E and Calabrese J 1991 *J. Solid State Chem.* **91** 121–125
[15] Baudelet F, Kong Q, Nataf L, Cafun J, Congeduti A, Monza A, Chagnot S and Itié J 2011 *High Pressure Res.* **31** 136–139
[16] Tetsuo I, Ayako K, Shizue S, Toru I and Hitoshi S 2003 *Nature* **421** 599–600
[17] Timoshenko J, Kuzmin A and Purans J 2014 *J. Phys.: Condens. Matter* **26** 055401

IOP Conf. Series: Materials Science and Engineering **503** (2019) 012019 doi:10.1088/1757-899X/503/1/012019

Nanoporous characterization of modified humidity-sensitive $MgO-Al_2O_3$ ceramics by positron annihilation lifetime spectroscopy method

H Klym[1,a], A Ingram[2], I Hadzaman[3], I Karbovnyk[4], I Vasylchyshyn[1] and A I Popov[5,b]

[1] Lviv Polytechnic National University, 12 Bandera St., Lviv 79646, Ukraine
[2] Opole University of Technology, 75 Ozimska St., Opole 45370, Poland
[3] Drohobych Ivan Franko State Pedagogical University, 24 I. Franko St., Drohobych 82100 Ukraine
[4] Ivan Franko National University of Lviv, 107 Tarnavskogo St., Lviv 79017, Ukraine
[5] Institute of Solid State Physics, University of Latvia, 8 Kengaraga St., Riga, LV-1063, Latvia

E-mail:[a]klymha@yahoo.com, [b]popov@latnet.lv

Abstract. Investigation of nanopores in the humidity-sensitive $MgO-Al_2O_3$ ceramics modified at 1100 °C for 2 hours was performed using positron annihilation lifetime spectroscopy method. Lifetimes of the third and fourth components of positron annihilation lifetime spectra obtained by four-component fitting procedure were used for calculation of nanopores radii. Transformation in free volumes was studied in ceramics after drying and exposure to water vapor. It is established that the lifetime of the third component decreases and the intensity of this component increases in ceramics exposed to water vapor reflecting ortho-positronium decay in nanopores simultaneously with annihilation in water bubbles. The characteristics of the fourth component (lifetime and intensity) decrease in ceramics exposed to water vapor reflecting ortho-positronium trapping in free-water volume of nanopores. Final drying of the $MgO-Al_2O_3$ ceramics results in returning of ortho-positronium lifetimes and intensities to the initial values confirming good desorption of water from nanopores of ceramics.

1. Introduction

Functional spinel $MgO-Al_2O_3$ ceramics are prospective nanoporous materials for humidity sensors [1-3]. It is known to have good chemical and thermal stability in comparison with other types of porous materials and short response time to humidity changes. As shown earlier [4,5], the functionality of humidity-sensitive ceramics is dependent on microstructure of grains, grain boundaries and open pores with different sizes. The latter significantly affect the functionality of humidity-sensitive ceramics.

Typically, microstructural properties of ceramic materials are studied using X-ray diffraction, electron microscopy, porosimetry methods etc. [6-11]. However, the methods of mercury and nitrogen porosimetry provide information on open pores with radius of >5 and > 2 nm, respectively [12,13]. Physical processes in ceramics depend not only on the number and nature of large open pores, but also on nanopores. Therefore, for more information on these structural components and their influence on the exploitation properties of technologically-modified $MgO-Al_2O_3$ ceramics, additional methods of structural characterization, which would allow to study free-volume entities on the nanoscale level are needed. In this case we used positron annihilation lifetime (PAL) spectroscopy technique to study free-volume transformation in spinel ceramics as an alternative to porosimetry methods [5]. As was shown earlier in [4,14], two channels of positron annihilation can be important in case of ceramics: positron trapping and ortho-positronium (o-Ps) decaying. The latter process (so called "pick-off" annihilation) resulting from Ps interaction with electron from environment (including annihilation in liquid water) is ended by emission of two γ-quanta. In general, these two channels of positron annihilation are independent. However, if trapping sites appear in a vicinity of grain boundaries neighboring with free-volume pores, these positron-Ps traps will become mutually interconnected resulting in a significant complication of PAL data [14].

In this work, characterization of nanopores in the $MgO\text{-}Al_2O_3$ ceramics obtained by annealing at 1100 °C for 2 hours are presented. O-Ps decay components extracted by four-component decomposition of positron annihilation lifetime spectra in ceramics are analyzed for different conditions such as drying and exposure to water vapor.

2. Experimental details

The studied $MgO\text{-}Al_2O_3$ ceramics was sintered at 1100 °C for 2 hours using traditional ceramic technology [15]. The PAL studies were performed using ORTEC spectrometer with ^{22}Na source at 22 °C and relative humidity of 35 %. Investigations were performed in several stages. The initial measurements were performed on ceramic samples dried in vacuum at 120 °C for 4 hours (after initial drying). To study the interaction of inner structure of nanoporous ceramic structures with water, the same samples were placed in a distiller at 100 % relative humidity for 7 days. Than ceramics were removed from distiller and placed in the ambient atmosphere at 22 °C and relative humidity of 35 %. PAL measurements were repeated on the 1st, 2nd, 3rd and 7th days after this procedure (after exposure to water vapor). Ultimately, the $MgO\text{-}Al_2O_3$ ceramic was dried in vacuum at 120 °C for 4 hours (after final drying) and measurements were repeated to determine PAL reversibility in the conditions of chemical and physical desorption of water molecules. In previous works [14,15], we used three-component fitting procedures with normal statistical treatment of PAL spectra that accumulate near one million of elementary positron annihilation events. At high-statistical measurements (more than ten millions of counts), the best results were obtained with four-term decomposition procedure. Such approach allows us to study nanopores of different sizes, responsible for o-Ps decaying. Each PAL spectrum was processed multiply owing to slight changes in the number of final channels, annihilation background and time shift of the 0-th channel. In such a manner, we obtained fitting parameters (positron lifetimes τ_1, τ_2, τ_3 and τ_4 and intensities I_1, I_2, I_3 and I_4), which correspond to annihilation of positrons in the samples of interest within a quite reliable error-bar using LT software [16]. The errors in the determination of lifetimes τ_i and intensities I_i are ±0.02÷0.2 ns (increasing with τ_i) and ±0.001÷0.01 a.u., respectively.

3. Results and Discussion

Typical PAL spectra of $MgO\text{-}Al_2O_3$ ceramics sintered at 1100 °C for 2 hours after initial drying and after exposure to water vapor are shown in Figure 1. As demonstrated earlier [15,17], in the case of humidity-sensitive $MgO\text{-}Al_2O_3$ ceramics, there are two channels of positron annihilation: positron trapping with τ_1 and τ_2 lifetimes and ortho-positronium o-Ps decaying within "pick-off" annihilation (τ_3 and τ_4 lifetimes). The first component describes microstructure of spinel, the second one corresponds to positron trapping in free-volume defects near grain boundaries. The third and the fourth components are due to "pick-off" annihilation of o-Ps in nanopores and in water molecules [17].

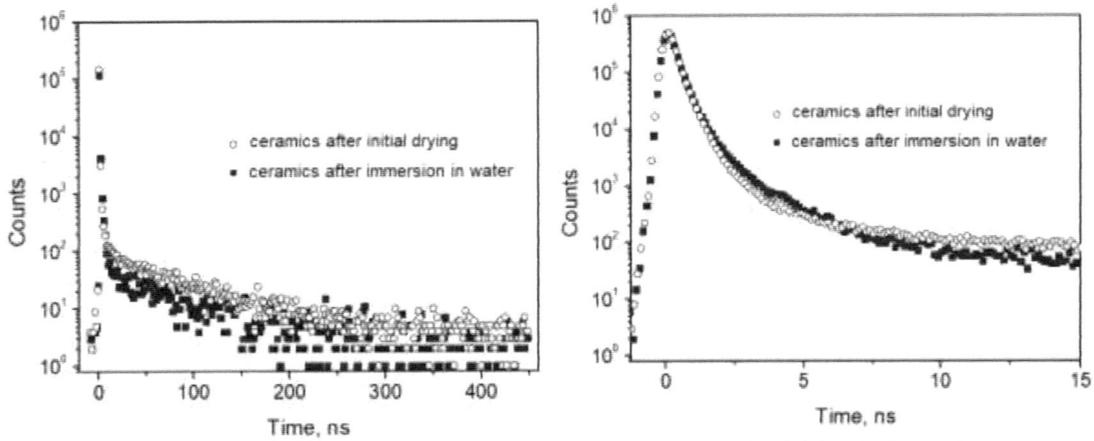

Figure 1. PAL spectra for $MgO-Al_2O_3$ ceramics sintered at 1100 °C after initial drying and exposure to water vapor.

As shown in Table 1, PAL parameters of the first component (τ_1,I_1) is practically unchanged since water adsorption processes do not affect structural features of spinels. The lifetime of the second defect-related component τ_2 varies around 0.46÷0.48 ns and the intensity I_2 is near 28 %. As a result, the positron trapping parameters calculated within two-state positron trapping model [15], such as average positron lifetime $\tau_{av.}$ and positron lifetime in defect-free bulk $\tau_{b.}$ remain unchanged, while positron trapping rate of defect κ_d slightly increases in ceramics after exposure to water vapor. Major changes caused by adsorption of water are observed in the third and fourth o-Ps-related components.

Table 1. Fitting parameters (for two components) and positron trapping modes of $MgO-Al_2O_3$ ceramics calculated within two-state positron trapping modes

Pre-history	Fitting parameters				Positron trapping modes		
	τ_1, ns	I_1, %	τ_2, ns	I_2, %	$\tau_{av.}$, ns	τ_b, ns	κ_d, ns^{-1}
After initial drying (ID)	0.169	68	0.462	28	0.254	0.21	1.10
1 day after exposure to water vapor (EW)	0.170	66	0.483	28	0.263	0.21	1.15
2 days after exposure to water vapor (EW)	0.170	65	0.457	29	0.258	0.21	1.14
3 days after exposure to water vapor (EW)	0.171	67	0.484	28	0.264	0.21	1.13
7 days after exposure to water vapor (EW)	0.167	67	0.462	29	0.257	0.21	1.16
After final drying (FD)	0.172	68	0.459	29	0.257	0.21	1.08

It is established that lifetimes of the third and the fourth components (τ_3 and τ_4) corresponds to the size of free volumes (nanopores), while intensities I_3 and I_4 – to the number of nanopores. As shown in Figure 2a, the lifetime τ_3 decreases in ceramics upon drying after exposure to water vapor. The intensity I_3 grows in ceramics after 1 day, indicating annihilation of o-Ps in water-filled nanopores. The presence of water in the nanopores is manifested by increase of intensity I_3 and a slight decrease of lifetime τ_3. It has to be noted that o-Ps lifetime of ~1,8 ns reflects the annihilation in the water bubbles [12,8].

The lifetime τ_4 and the intensity I_4 significantly decrease in ceramics after 1 day (Figure 2b). It relates to the penetration of nanopores by water molecules (full or partial filling) changing characteristics of o-Ps-related component. Obviously, such nanopores should have access to environmental and internal communications at the nanoscale. The intensity I_4 does not fall to zero and is near 0.9 %, it should be assumed that part of the nanopores where the disintegration of o-Ps atoms occurs, is closed. Naturally, the grains growth and the formation of closed porosity begins at the nanoscale. PAL method also provides information on the phase separation during the early stages, which contributes to a closed porosity. The third and fourth components (τ_3, I_3) and (τ_4, I_4) originate from annihilation of o-Ps atoms in intrinsic nanopores (closed and open) of MgO-Al_2O_3 ceramics. But only open nanopores can contain adsorbed water [14].

Additional investigation of the MgO-Al_2O_3 ceramics on the 2nd, 3rd and 7th day shows the increase of the lifetime τ_4 and the intensity I_4, indicating release of water from the inner voids of ceramics. After final drying in vacuum at 120 °C for 4 hours, the initial distribution of nanopores in MgO-Al_2O_3 ceramics tends to recover (Figure 2b). However, the parameters of the fourth component are not fully recovered, proving that some water molecules remain adsorbed.

The intensity I_3 decreases after 2, 3 and 7 days and after final drying confirms good desorption ability of studied MgO-Al_2O_3 ceramics (Figure 2a). Reducing the value of the lifetime τ_4 after final drying of ceramics sintered at 1100 °C with poorly developed porosity [4] can be due to the formation of thin layers of water molecules surrounding larger nanopores.

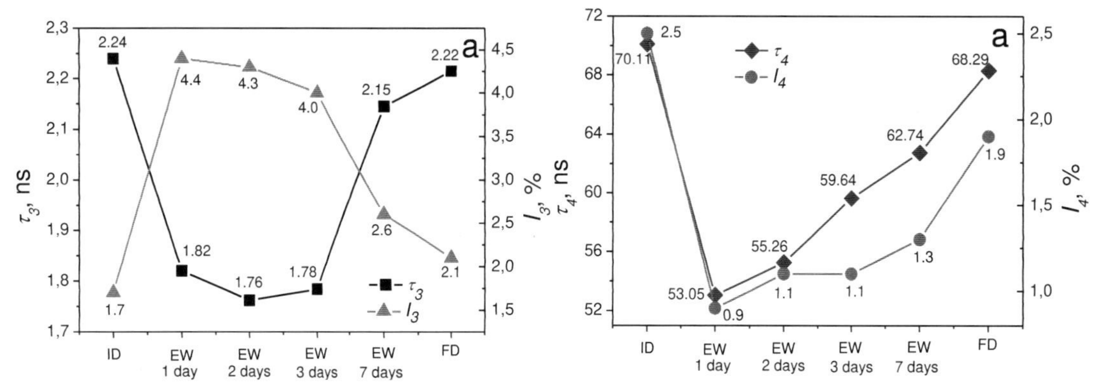

Figure 2. Dependences of the lifetime τ_3 and the intensity I_3 (a) as well as the lifetime τ_4 and the intensity I_4 (b) on the conditions of watering and drying: ID – after initial drying, EW – after exposure to water vapor, FD – after final drying.

The lifetimes τ_3 and τ_4 are used for calculation of radii (R_3 and R_4) of nanopores within Tao-Eldrup model [19-21]:

$$\tau_{o-Ps} = \left[2\left(1 - \frac{R}{R+\Delta R} + \frac{1}{2\pi}\sin\left(\frac{2\pi R}{R+\Delta R} \right) \right) + 0.007 \right]^{-1}, \quad (1)$$

where ΔR is the empirically determined parameter (in the classical case $\Delta R \approx 0.1656$ nm), describing effective thickness of the electron layer responsible for the "pick-off" annihilation of *o-Ps* in the hole.

The contribution corresponding to nanopores with radii R_3 and R_4 was calculated by semiempirical equation:

$$f_v = C \cdot V_f \cdot I_{o-Ps}, \quad (2)$$

where $V_f = 4/3 \cdot \pi \cdot R_{o-Ps}$ – free volume calculated from o-Ps-related components; I_{o-Ps} – intensities of o-Ps-related components, C – empirical parameter (0,0018).

Changes of radius of free volumes (or size of nanopores where o-Ps decayed) caused by water sorption processes in the MgO-Al_2O_3 ceramics, sintered at 1100 °C for 2 hours are shown in Figure 3a. The most significant changes in ceramics caused by immersion of water are observed in largest nanopores with radius of ~18 Å. The intensity of this component (as well as intensity I_3) does not return to initial value after final drying of ceramics (Figure 3a and Figure 3b), because not all water is eventually released from nanopores. As was shown in [14,15], the o-Ps "pick-off" annihilation in nanopores with radius R_3 occurs in dry pores and in the "bubbles" of liquid water adsorbed by these pores. The o-Ps "pick-off" annihilation in nanopores with radius R_4 occurs in free volume of pores, because physisorbed water molecules are formed at the pore walls in the form of thin layer.

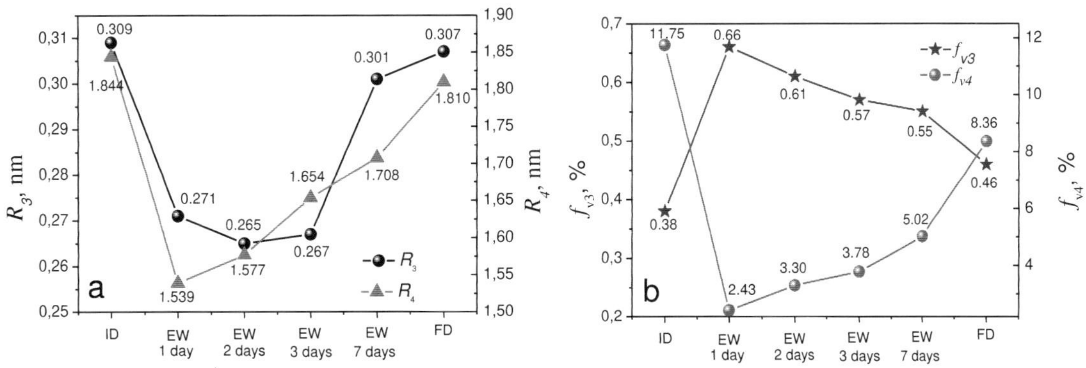

Figure 3. Dependences of radius R_3 and R_4 (a) as well as f_3 and f_4 (b) on the conditions of watering and drying: ID – after initial drying, EW – after exposure to water vapor, FD – after final drying.

4. Conclusions

It is established that the lifetimes of the third and the fourth o-Ps-related components of PAL spectra decrease in MgO-Al_2O_3 ceramics after exposure to water vapor. The free volume of the largest nanopores decreases, while o-Ps decay in smaller nanopores occurs simultaneously with "pick-off" annihilation in water bubbles. The Ps annihilation in nanopores with adsorbed water vapor is shown to occur via two mechanisms: o-Ps decaying in nanopores with radius of 0.3 nm including "pick-off" annihilation in liquid water and o-Ps trapping in free volume of nanopores with physisorbed water molecules at the pore walls.

Acknowledgments

The work was supported by the Ukrainian Ministry of Education and Science. H. Klym thanks Prof. O. Shpotyuk for the fruitful discussion.

References

[1] Lee C Y and Lee G B 2005 *Sensor Letters* **3** 1.
[2] Laobuthee A, Wongkasemjit S, Traversa E and Laine. 2000 *J. Europ. Ceram. Soc.* **20** 91
[3] Luo X, Zheng X, Duan J Xu J, Yang Y and Xiao Z 2003 *J. Chinese Ceram. Soc.* **31** 1058
[4] Klym H, Hadzaman I, Shpotyuk O 2015 *Materials Science* **21** 92-5
[5] Klym H, Ingram A, Shpotyuk O, Hadzaman I, Solntsev V, Hotra O and Popov A I 2016 *Low Temperature Physics* **42** 601
[6] Popov A I, Shirmane L, Pankratov V, Lushchik A, Kotlov A, Serga V E, Kulikova L D, Chikvaidze G and Zimmermann J 2013 *Nucl. Instrum. Meth. B* **310** 23
[7] Karbovnyk I, Borshchyshyn I, Vakhula Y, Lutsyuk I, Klym H and Bolesta I 2016 *Ceramics International* **42** 8501
[8] Lushchik A, Dolgov S, Feldbach E Pareja R, Popov A I, Shablonin E and Seeman V 2018 *Nucl. Instrum. Meth. B* **435** 31
[9] Vieira C M F, da Silva P R N, da Silva F T Capitaneo J L and Monteiro S N 2005 *Revista Matéria* **10** 526
[10] Fukasawa T, Ando M, Ohji T and Kanzaki S 2001 *J. Amer. Ceram. Soc.*, **84** 230
[11] Ganesh I, Johnson R, Rao G V N Mahajan I R, Madavendra S S and Reddy B M 2005 *Ceramics International* **31** 67
[12] Golovchak R, Wang S, Jain H and Ingram A 2012 *J. Mater. Research* **27** 2561
[13] Gregorová E, Živcová Z and Pabst W 2009 *Starch* **61** 495
[14] Klym H, Ingram A, Shpotyuk O, Hadzaman I and Solntsev V 2016 *Nanoscale research let.* **11** 133
[15] Klym H, Ingram A, Shpotyuk O, Hadzaman I, Hotra O and Kostiv Yu 2016 *J. Mater. Eng. Perform.* **25** 866
[16] Kansy J 1996 *Nucl. Instrum. Meth. Phys. Res. A* **374** 235.

[17] Klym H, Ingram A and Shpotyuk O 2010 *Proc 27th Internat.Conf. Microelectronics (MIEL)*, 239.
[18] Kotera K, Saito T, Yamanaka T 2005 *Physics Letters A* **345** 184
[19] Goworek T. 2002 *Chem. Phys. Let.* **366** 184
[20] Rigby S P, Fletcher R S, Riley S N 2004 *Chemical Engineering Science* **59** 41
[21] Wang J, Lin Q, Zhou R and Xu B 2002 *Sensors and Actuators B: Chemical* **81** 248

IOP Conf. Series: Materials Science and Engineering **503** (2019) 012020 doi:10.1088/1757-899X/503/1/012020

Structural investigation of crystallized Ge-Ga-Se chalcogenide glasses

H Klym[1,a], O Shpotyuk[2,3], I Karbovnyk[4], L Calvez[5] and A I Popov[6,b]

[1] Lviv Polytechnic National University, 12 Bandera St., Lviv 79646, Ukraine
[2] Vlokh Institute of Physical Optics, 23 Dragomanova St., Lviv 79005, Ukraine
[3] Jan Dlugosz University in Czestochowa, 13/15, Armii Krajowej St., Czestochowa 42200, Poland
[4] Ivan Franko National University of Lviv, 107 Tarnavskogo St., Lviv 79017, Ukraine
[5] Equipe Verres et et Céramiques, UMR-CNRS 6226, Institute des Sciences chimiques de Rennes, Université de Rennes 1, Rennes Cedex, France
[6] Institute of Solid State Physics, University of Latvia, 8 Kengaraga St., Riga, LV-1063, Latvia

E-mail:[a]klymha@yahoo.com, [b]popov@latnet.lv

Abstract. Crystallization transformation in the 80$GeSe_2$-20Ga_2Se_3 chalcogenide glasses caused by annealing at 380 °C during different duration (25, 50, 80 and 100 hours) are studied using X-ray diffraction and atomic force microscopy methods. It is established that $GeGa_4Se$ phase of low- and high-temperature modification, Ga_2Se_3 phase (α- and γ-modification) and $GeSe_2$ phases are crystallized during this process. It is shown that annealing duration over 50 h does not lead to further internal structural crystallization, while annealing for 80 h result in processes of surface crystallization.

1. Introduction

Development of modern IR photonics brings up the challenges of searching new functional media for effective transfer of wide-spectra electro-magnetic radiation and developing of novel miniaturization technologies of passive and active photonic elements (optical waveguides, resonators, splitters, multiplexers, detectors, signal amplifiers and converters, comparators etc). Among the most promising materials for such applications are special chalcogenide glasses (ChG) – non-oxide glassy-like materials with high content of chalcogens (S, Se, Te) [1-3]. To a great extent, the further success in this field relies on chemical-technological resolutions in development of ChG-based media with unique properties [4-6].

Basic approach in development of functional materials is based on the various methods of technological and post-technological structural modification using external factors such as thermal annealing, high-energy radiation treatment or laser beams [7-9]. It is known that the nearest atomic environment in glass matrix can be adequately studied using numerous experimental methods such as vibration and Raman scattering spectroscopy, X-ray diffraction (XRD), scanning electron microscopy etc. [10].

The aim of this work is the investigation of crystallization processes and features formation of crystalline phases in 80$GeSe_2$-20Ga_2Se_3 chalcogenide glasses under different durations of thermal annealing above the glass transition temperature using XRD and atomic force microscopy (AFM) methods.

2. Experimental details

The $80GeSe_2$-$20Ga_2Se_3$ samples were obtained from high purity (99.999%) Ge, Ga and Se components in silica ampoule kept under 10^{-6} vacuum. Raw materials were heated from 20 to 850 °C at 2 °C/min and maintained at the highest temperature for 12 h. Then silica tube was cooled in water at room temperature and annealed at 30 ° C below the glass transition temperature T_g (370 °C) for 3 h and slowly cooled down to room temperature [11]. Samples with thickness of 1 mm were polished for further investigations [9, 12-14].

The crystallization of the $80GeSe_2$-$20Ga_2Se_3$ ChG was performed with a single step typical heat treatment at T_g+10 °C as described in [11,15]. This temperature has been chosen as an optimal temperature of ceramization as it permits to control the generation by simultaneous nucleation and growth of nanoparticles within the glassy matrix according to the heat treatment time. Thus, glass samples were placed in a ventilated furnace where the accuracy of temperature is ± 2 °C for various time varying from 25 to 100 h [12].

The XRD measurements ($CuK_{\alpha 1}$ radiation) were carried out to determine crystalline phases in the studied glasses. The automatic STOE STADI P diffractometer ("STOE & Cie GmbH, Germany) with a linear position-precision detector was used for XRD measurements. All measurements were conducted in 2θ-step regime, the profiles of peaks being refined using WinPLOTR software [12,15].

The surface morphology of the glasses annealed at 25 and 80 h was studied by means of Solver P47-PRO AFM, the obtained images being processed with Image Analysis program (NT-MDT).

3. Results and Discussion

It is possible to assume that basic transformations can be related to inwardly-phase stratification in the $80GeSe_2$-$20Ga_2Se_3$ cut-section on individual components (stoichiometric $GeSe_2$ and Ga_2Se_3) which set the stage for further reaction forming of $GeGa_4Se_8$ triple phase [12]. This indicates that the stoichiometric $80GeSe_2$-$20Ga_2Se_3$ cut-section creates terms for crystallization of $GeGa_4Se_8$ phase due to segregation of preliminary selected $GeSe_2$ and Ga_2Se_3 phases with additional formation of $GeSe_2$-enriched residues. Since the stoichiometric $GeSe_2$ phase is outside glass formation [12] then in future it will be formed in a separate crystalline phase.

The Ge-Ga-Se system (polythermal and isothermal cut-sections, crystal structure of intermediate phase) was studied in [16]. $GeGa_4Se_8$ ($Ga_{0.5}Ge_{0.13}Se$) phase exists in two modifications with polymorphic transformation temperature of ~898 K (625 °C). High-temperature modification is crystallized in ZnS structural type (cubic crystal system, F-43m space group, a ~5,45 Å), whereas for low-temperature modification ($GeGa_2Se_5$) structure has not been established exactly, however, lattice parameter a is known to be ~ 5,461 Å [16] indicating affinity in structures of these two modifications. It was also reported about existence of $GeGa_2Se_3$ but $GeGaSe_3$ phases [17]. Total structural relationship between these phases and binary Ga_2Se_3 $GeSe_2$ phases and various modifications are Ga [Se_4] and Ge [Se_4] tetrahedrons.

Combined results obtained by XRD method for $80GeSe_2$-$20Ga_2Se_3$ chalcogenide glasses before and after thermal annealing at 380 °C for 25, 50, 80 and 100 h are shown in Figure 1. In the $80GeSe_2$-$20Ga_2Se_3$ ChG annealed for 25 h and 50 h the well-formed peaks of $GeGa_4Se_8$, Ga_2Se_3 and $GeSe_2$ phases appear. The maximal reflection corresponding to $GeSe_2$ phase is semi-amorphous halo while reflections from $GeGa_4Se_8$ and Ga_2Se_3 are relatively well formed, especially at maximum of $2\theta \sim 28.22°$ (Figure 1). The intensity at low diffraction angles of 2θ is shown suggesting that the fractal formation in the glass during annealing will not disappear, but become larger.

This process is accompanied by crystallization of $GeGa_4Se_8$ phase (in the structure of glass from chaotic arrangement of Ga [Se_4] and Ge [Se_4] tetrahedrons is characterized by only short-range order, the transition to the formation of long-range order characteristic of crystalline structure of $GeGa_4Se_8$ with ZnS structural type ZnS occurs).

Further increasing in annealing duration to 80 and 100 h does not affect the diffraction peaks, reflecting mainly the formation of double and triple $GeGa_4Se_8$ and Ga_2Se_3 phases, just as for $80GeSe_2$-$20Ga_2Se_3$ glass, annealed for 50 h (Figure 1). Thus, further crystallization process registered by XRD and formation of long-range order during such annealing do not occur. The first maximum of semi-amorphous halo at ~14.6 °2θ corresponds to the maximum intensity reflection (002) of $GeSe_2$ phase (monoclinic syngony, $P2_1/c$ space group).

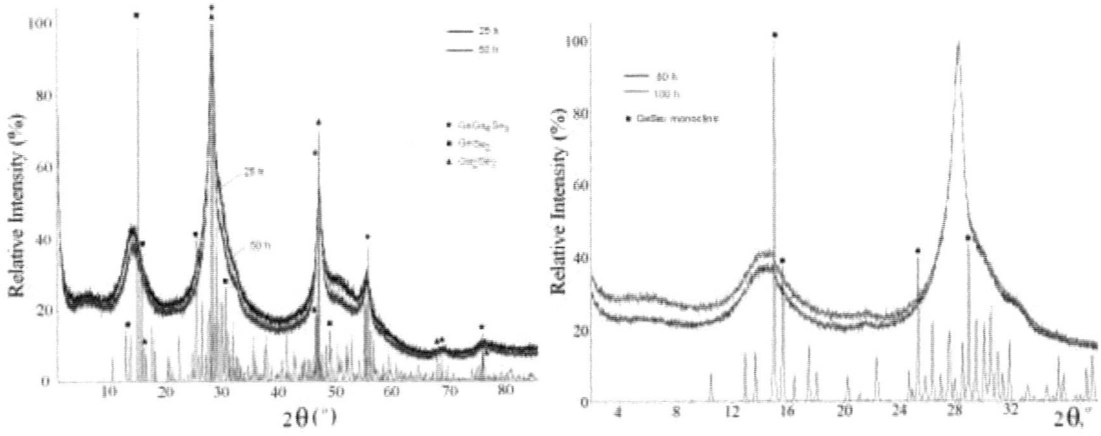

Figure 1. Experimental XRD patterns for initial 80GeSe2-20Ga2Se3 glasses and after annealing for 25, 50, 80 and 100 h: comparison of experimental results with theoretical lines.

For ChG samples annealed at high duration of 80 and 100 h the additional analysis of Ga_4GeSe_8 and Ga_2Se_3 crystallization phase was performed. Considering the state diagram for the Ga-Se system, temperature of existence of cubic Ga_2Se_3 (high-temperature $Ga_{0.67}Se$) begins above $T = 730$ °C. Below this temperature low-temperature monoclinic modification of Ga_2Se_3 exists. Distinguishing of monoclinic and cubic modification of Ga_2Se_3 phases is possible only on samples with the high degree of crystallization.

Assuming that in studied samples pure high-temperature Ga_2Se_3 phase is crystallized, most likely this is modification of α-Ga_2Se_3 ($a \sim 5.44$ Å). However, taking into account the fact that the peaks are extended, along with α-modification γ-modification of Ga_2Se is possible (Figure 2). Yet, one should pay attention to the raising of the background on diffraction pattern (indicated by ellipses in Figure 3).

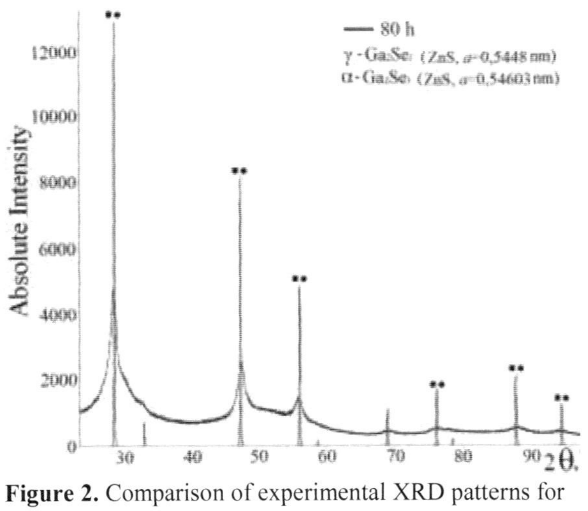

Figure 2. Comparison of experimental XRD patterns for ChG annealed for 80 h with theoretical reflexes of α-Ga_2Se_3 and γ-Ga_2Se_3 phases

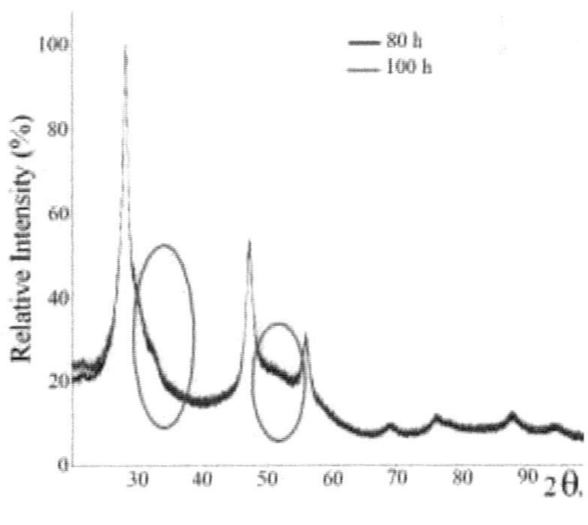

Figure 3. Raising of the background on experimental XRD patterns for samples annealed for 80 and 100 h.

These raising may correspond to peaks from experimental diffraction pattern for high- and low-temperature modification of Ga_4GeSe_8 (Figure 4) but we can also assume that crystallized high-temperature α-Ga_2Se_3 phase may exist. Amorphous phase of glass is likely to be responsible for the features on the pattern in Figure 4 outlined by ellipses.

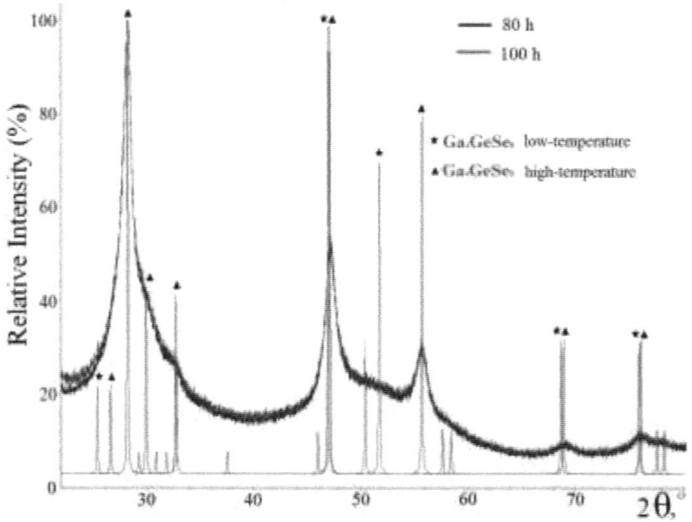

Figure 4. Comparison of experimental XRD patterns for ChG annealed for 80 and 100 h with theoretical reflexes of Ga_4GeSe8 phase low- and high-temperature modifications.

The width of the peak at 2θ ~ 28,22° indicates the presence of nanoparticles dispersed in the form of nanocrystalline inclusions with the size of 9-10 nm (determined by the Debye-Scherrer equation [12,18]) in the crystal matrix. It should be underlined that the height of this peak in glasses annealed at 80 and 100 h does not change essentially as compared to ChG treated at 50 h. Such behavior speaks in favor of saturated crystallization at longer durations of annealing.

The maxima associated with $GeSe_2$ phase appear on the XRD patterns of thermally annealed $80GeSe_2$-$20Ga_2Se_3$ glass as well [14] but (in contrast to [19]) they cannot be well distinguished as separate crystalline peaks even for prolonged annealing. It means that $GeSe_2$ crystals appear only in a small amount.

To better understand these processes on the surface, the glasses were examined by AFM (Figure 5).

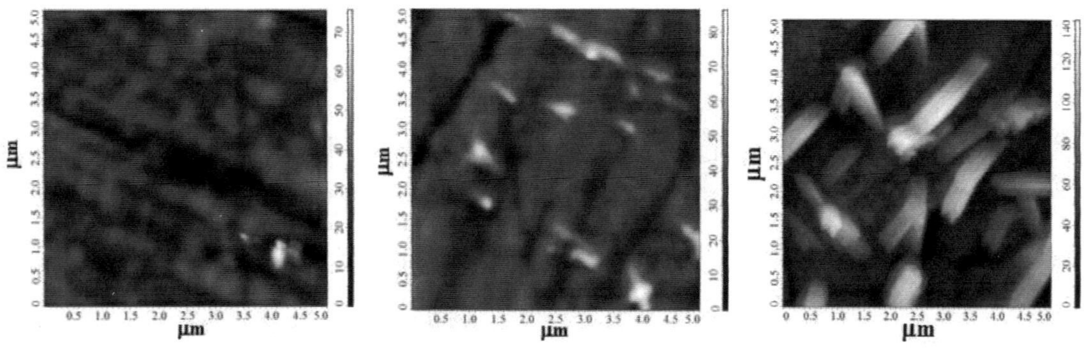

a b c

Figure. 5. AFM images of unannealed $80GeSe_2$-$20Ga_2Se_3$ glasses (a) and annealed at 380 °C for 25 h (b) and 80 h (c).

As was shown in [15,20], that surface of base unannealed $80GeSe_2$-$20Ga_2Se_3$ glass is mostly morphologically uniform (Figure 5, a). At various cycles of potential scan, surface irregularities and individual features appear. It can be caused by hitting with microparticles from air since microscope stand is not placed in a vacuum chamber. The elemental analysis of the surface testifies stoichiometric composition of $Ge_{23,5}Ga_{11,8}Se_{64,7}$ glass [21-23]. However, the thermal annealing of glasses for 25 h causes obvious changes in the surface morphology (Figure 5, b). After analysing the images, we can conclude that thermal annealing at 380 °C for small durations leads to internal transformation of free volume in the $80GeSe_2$-$20Ga_2Se_3$ glass due to the formation of additional crystalline phase on the surface [15,20,24]. However, after longer treatment (over 50 h), surface crystallization occurs more efficiently. As shown in Figure 5, crystallization of $GeSe_2$ phase in samples annealed for 80 h at 380 °C is a surface phenomenon. With respect to AFM and SEM data [15], the $GeSe_2$ crystals in form of wires with 1-3 μm lengths are non-uniformly distributed on sample surface (Figure 5,c).

4. Conclusions

It is established that crystallization processes in the $80GeSe_2$-$20Ga_2Se_3$ ChG caused by annealing at 380 °C for 25, 50, 80 and 100 h indicate the formation of $GeGa_4Se$ (high- and low-temperature modification), Ga_2Se_3 (α- and γ-modification) and $GeSe_2$ crystals with size of 9-10 nm. Increasing duration of thermal treatment for $80GeSe_2$-$20Ga_2Se_3$ glasses leads to obtaining of thermally stable glasses and glass-ceramic media. Longer annealing durations result in surface crystallization of $GeSe_2$ phase.

Acknowledgment

H. Klym thanks to the Ministry of Education and Science of Ukraine for support and Dr. P. Demchenko for the assistance in XRD experiments.

References

[1] Calvez L, Ma HL, Lucas J and Zhang X-H 2007 *Advanced materials* **19** 129
[2] Lezal D 2003 *J. Optoelectron. Adv. Mater* **5** 23
[3] Lezal D, Pedlikova J and Zavadil J 2004 *J. Optoelectron. Adv. Mater* **6** 133
[4] Rozé M, Calvez L, Ledemi Y, Allix M, Matzen G and Zhang X-H 2008 *J. Amer. Ceram. Soc.* **91** 3566
[5] Sanghera J S and Aggarwal I D 1999 *J. Non-Crystalline Solids* **256** 6
[6] Owen A E, Firth A P and Ewen P J S 1985 *Philosophical Magazine B* **52** 347
[7] Monge M A, González R, Muñoz Santiuste J E, Pareja R, Chen Y, Kotomin E A and Popov A I 2000 *Nucl. Instrum. Meth. B* **166-167** 220
[8] Popov A I and Balanzat E 2000 *Nucl. Instrum. Meth. B* **166-167**, 545-9
[9] Karbovnyk I, Olenych I, Aksimentyeva O, Klym H, Dzendzelyuk O, Olenych Yu and Hrushetska O 2016 *Nanoscale research letters* **11** 84
[10] Calvez L, Lin C, Rozé M, Ledemi Y, Guillevic E. Bureau B, Allix M and Zhang X-H 2010 *Proc. SPIE* **7598** 759802-1
[11] Li J, Wang G, Lin C, Zhang T, Zhang R, Huang Z, Shen X, Gu B, Ye B, Ying F, Li M and Nie Q *Infrared Physics & Technology* 2017 **83** 238-42
[12] Shpotyuk O, Calvez L, Petracovschi E, Klym H, Ingram A and Demchenko P 2014 *J. Alloys&Compounds* **582** 323
[13] Klym H, Ingram A and Shpotyuk O 2016 *Materialwissenschaft und Werkstofftechnik* **47**(2-3), 198.
[14] Klym H, Ingram A, Shpotyuk O, Hotra O and Popov A I 2016 *Radiation Measurements* **90** 117
[15] Klym H, Ingram A, Shpotyuk O, Calvez L, Petracovschi L, Kulyk B, Serkiz R and Szatanik R 2015 *Nanoscale research letters* **10** 49
[16] Thiebault C, Guen L, Eholie R, Flahaut J and Ollitrault-Fichet R 1975 *Bull. Soc. Chim. France* **5–6**, 967
[17] Rouland J C, Ollitrault-Fichet R, Flahaut J, Rivet J and Ceolin R 1990 *Thermochimica Acta* **161** 189
[18] Keshari A K and Pandey A C 2008 *Journal of Nanoscience and Nanotechnology* **8** 1221
[19] Rozé M, Calvez L, Hubert M, Toupin P, Bureau B, Boussard-Plédel C and Zhang X-H 2011 *Intern. J. Applied Glass Science* **2** 129

[20] Klym H, Karbovnyk I, Kostiv Y and Vasylchyshyn I 2018 *Proc. 38th Internat. Conf. Electronics and Nanotechnology (ELNANO)*, 152.

[21] Munzar M, Koughia K, Tonchev D, Kasap S O, Sakai T, Maeda K, Ikari T, Haugen C, Decorby R and McMullin J N 2006 *Phys. Chem. Glasses* **46** 215

[22] Calvez L, Novikova A, Potapov A and Velmuzhov A 2017 *Optical Components and Materials XIV* **10100** 1010014

[23] Tang Z, Neate N C, Furniss D, Sujecki S, Benson T M and Seddon A B 2011 *J. Non-Crystalline Solids* **357** 2453

[24] Hubert M, Delaizir G, Monnier J, Godart C, Ma H-L, Zhang X-H and Calvez L 2011 *Optic Express* **19** 23513

Luminescence characteristics of magnesium aluminate spinel crystals of different stoichiometry

G Prieditis[1,a], E Feldbach[1], I Kudryavtseva[1], A I Popov[2b], E Shablonin[1] and A Lushchik[1,c]

[1]Institute of Physics, University of Tartu, W. Ostwald 1, 50411 Tartu, Estonia
[2]Institute of Solid State Physics, University of Latvia, Kengaraga 8, Riga LV-1063, Latvia

E-mail: [a]prieditisgatis@gmail.com, [b]popov@latnet.lv, [c]aleksandr.lushchik@ut.ee

Abstract. Magnesium aluminate spinel single crystals with different stoichiometry, $MgAl_2O_4$ (1:1 spinel) and $MgO{\cdot}2.5Al_2O_3$ (1:2.5) were investigated using different optical methods (cathode-, photo- and thermally stimulated luminescence (TSL), optical absorption, "creation spectra" of TSL peaks and phosphorescence by VUV radiation). Low-temperature charge carrier traps and the position of intrinsic UV emission bands depend on the degree of stoichiometry. Antisite defects (ADs), Mg^{2+} or Al^{3+} located in a "wrong" cation site ($Mg|_{Al}$ or $Al|_{Mg}$) are the main as-grown structural defects, which serve also as efficient traps for electrons and holes as well as seeds for bound excitons. AD concentration is especially high in 1:2.5 spinel. There are several manifestations of ADs (electronic excitations near ADs) in the spectral region of 7–7.5 eV, slightly below the energy gap.

1. Introduction

Magnesium aluminate spinel is a fascinating material applicable for various applications in science and technology [1]. Its particular feature is a very high tolerance against radiation and absence of swelling even after heavy fast-neutron irradiation (see, e.g., [2,3]). Therefore, $MgAl_2O_4$ crystals and transparent ceramics are considered as attractive candidates for diagnostics/optical windows in future fusion projects (ITER, DEMO, PROTO, industrial reactors) [4]. It is generally accepted that the functionality of optical components strongly depends on the concentration of as-grown structural defects and their accumulation during material exposure to dense and prolonged irradiation.

Mg-Al spinel is a mixed double oxide $MgO{\cdot}nAl_2O_3$ and may be regarded as a face-centered cubic structure formed by closely packed oxygen ions. In equimolar mixture of binary oxides – stoichiometric $MgAl_2O_4$ (melting temperature 2400 K), Mg^{2+} ions occupy every eighth tetrahedral (T) interstice and a half of octahedral (O) ones are filled by Al^{3+} in each unit cell. In a "normal" $MgAl_2O_4$, there is no cation disorder, while the exchanging of cations between T and O sites results in the formation of so-called antisite defects (ADs) – $Mg|_{Al}$ or $Al|_{Mg}$ – a cation in a "wrong" position. The formation energy of these ADs in $MgAl_2O_4$ is much lower than the one for all other point defects [5].

The presence of ADs charged -1 or +1 with respect to a regular lattice determines the degree of cation disorder ascribed by inversion parameter i. In a fully inversed spinel ($i = 1$), all Mg^{2+} are located in Al^{3+} sites, i.e. are ADs. In nonstoichiometric spinel, for instance $MgO{\cdot}2.5Al_2O_3$ Al-rich crystals studied in the present work, some Al^{3+} replace Mg^{2+} with the creation of positively charged $Al|_{Mg}$, while additional cation vacancies (mainly *O*-coordinated, i.e. aluminium vacancies V_{Al}) are formed for charge compensation [6].

There are many experimental and modelling studies related to the structural damage of radiation-tolerant Mg-Al spinel crystals exposed to dense irradiation of different types – high-energy electrons, fast neutrons, swift heavy ions (see [3,5,7-12] and references therein). On the other hand, the concentration of as-grown structural defects (mainly ADs or AD-containing complex defects) in the majority of Mg-Al single crystals is rather high; *i* reaches 0.1–0.2 in our 1:1 samples (determined by XRD in accordance with [13]) and is even higher in transparent optical ceramics. Therefore, the present study is devoted to the optical characterization of as-grown structural defects in stoichiometric and nonstoichiometric Mg-Al spinel crystals.

2. Experimental details

Single crystals of stoichiometric $MgAl_2O_4$ (labelled as 1:1 spinel) were grown by Union Carbide Corporation using the Czochralski method and nonstoichiometric $MgO{\cdot}2.5Al_2O_3$ samples (1:2.5 spinel) were grown using the Verneuil method. Both samples contained Cr, Mn and Fe impurities.

Cathodoluminescence (CL) measurements (see for details [14,15]) were performed using an electron gun operating at 10 keV and 2.5 μA and a close-cycle He cryostat. CL was analyzed using vacuum double monochromator built up at the Institute of Physics, Tartu. After the electron irradiation was stopped, it was possible to register the spectra of phosphorescence at 6 K and the thermally stimulated luminescence (TSL) at 6–420 K (heating rate of $\beta = 10$ K min^{-1}) for an integral signal or a specific monochromator-selected emission.

The spectra of optical absorption in UV-VUV spectral region were measured at room temperature (RT) using a homemade setup based on a single vacuum monochromator VMR-2. The number of incident photons from the hydrogen discharge in a flow capillary tube was maintained constant by varying the monochromator slit width and using the constant signal from sodium salicylate for normalization. The "creation spectra" of several TSL peaks ($\beta = 10$ K min^{-1}, 80–450 K) and phosphorescence by selective VUV radiation (prescribed number of exciting photons at each of several energies) were measured using the same setup and highly sensitive methods elaborated in Tartu earlier (see, e.g., [16,17]).

For photoluminescence (PL) measurements, a Hamamatsu deuterium discharge lamp was used as excitation source, McPherson vacuum and ANDOR Shamrock monochromators, a photon counting head and close-cycle He cryostat. The excitation spectra were normalized to equal the quantum intensities of the radiation falling onto the crystal (a sodium salicylate was used as a reference signal).

3. Results and Discussion

Figure 1*a* presents the spectra of CL measured for 1:1 and 1:2.5 spinel single crystals at various temperatures. The general spectrum shape is similar, but the spectra still differ, to some extent, in stoichiometric and nonstoichiometric crystals. At 6–78 K, the CL is more intense (by more than five times for UV luminescence band) in the 1:1 spinel, and the attenuation of CL, especially in UV spectral region, takes place with the temperature rise to RT.

Based on the literature data ([12,18–23] and references therein), the emission bands peaked at ~1.8 and ~2.4 eV can be ascribed to Cr^{3+} and Mn^{2+} impurity ions, respectively, while some of UV bands are of intrinsic origin partly related to ADs (excitonic type and recombination luminescence). Based on the similarity to MgO and Al_2O_3 crystals being constituent parts of spinel, the emission bands around 3 eV can be partly related to F and F^+ centers (two or one electron in the field of an oxygen vacancy V_O). These F- and F^+-emissions should be well pronounced in the crystals exposed to dense irradiation providing knock-on (impact) creation of novel F and F^+. Nevertheless, in as-grown (virgin) spinel crystals, there can exist a certain amount of V_O, and the emission of F^+ and/or F centers appears due to

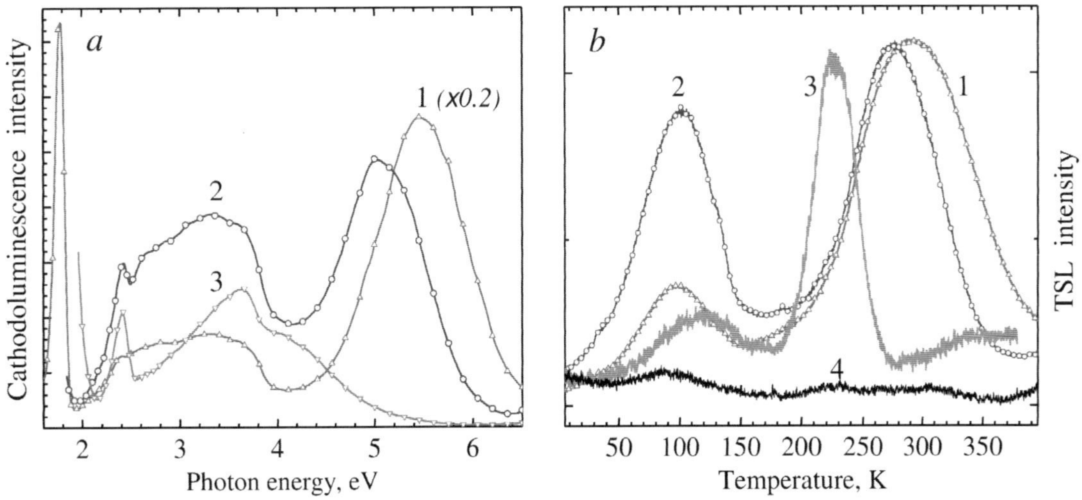

Figure 1. *a* – Cathodoluminescence spectra measured for virgin $MgAl_2O_4$ (curve 1) and $MgO{\cdot}2.5Al_2O_3$ (2 and 3) single crystals under excitation by 10 keV electrons at 78 K (1), 6 K (2) and RT (3). *b* – TSL curves measured after irradiation of virgin 1:1 (curves 1 and 2); virgin 1:2.5 (curve 3) and neutron-irradiated 1:2.5 (curve 4) samples by an10-keV electron beam at 6 K. $\beta = 10$ K min^{-1}.

the interaction of electrons and holes (*e-h* pairs), formed under electron beam excitation, with V_O (for details see, e.g., [15, 24]). To the best of our knowledge, up to now only the ~2.7-eV emission is reliably interpreted as the F^+-center emission [21], which is relatively intense in our 1:2.5 spinel crystal.

The presence of as-grown structural defects can be also visualized measuring TSL of the crystals exposed to 10-keV electron irradiation, i.e. after prolonged CL measurements. Electrons of such energy are not able to form novel structural defects, while already existing as-grown structural defects (as well as some impurities) serve as efficient electron/hole traps for *e-h* formed by an electron beam. Figure 1*b* shows the normalized TSL curves measured at a heating of the 1:1 and 1:2.5 spinel single crystals previously irradiated by 10-keV electrons at 6 K. It is clearly seen that different TSL peaks dominate in two 1:1 samples from different suppliers and the 1:2.5 crystal. This circumstance testifies to the different type/concentration of as-grown structural defects (incl. charged ADs - $Mg|_{Al}$ and $Al|_{Mg}$) in the crystals with different stoichiometry. The analysis of the detected TSL peaks is still in the progress. The first order kinetics gives rather "strange" results due to complex and unusual origin of the TSL peaks in spinel crystals with high level of structural disorder (see also text related to Figure 3). Note that TSL in a fast-neutron irradiated 1:2.5 crystal measured under the same conditions is strongly suppressed. The same situation occurs if we compare CL spectra for the virgin and neutron irradiated crystals (see also [12,15]). Such luminescence attenuation is mainly caused by CL reabsorption by neutron-induced defects (e.g., F and F^+ centers) but the contribution of radiation-induced enhancement of lattice inversion cannot be excluded as well.

A set of emission spectra was measured at the photoexcitation (from 6 to 8.1 eV) of different spinel samples at RT or 6 K. As an example, Figure 2*a* presents some PL spectra for two 1:1 spinel samples at 6 K. There is a certain similarity with the CL spectra presented in Figure 1*a*. At RT, the intense phosphorescence (PhP), detectable after the excitation had been stopped, was registered in the sample preliminary excited with photons of $h\nu >$ ~7.5 eV (see text related to Figure 3*b*). The PhP spectrum for the 1:1 crystal excited by 8.1-eV photons is presented in Figure 2*a* (curve 3). Such intense PhP disturbs the spectra of PL excited by photons from the region of fundamental absorption and artificially increases the PL intensity in UV region (see curve 1′*a*). At helium temperatures, PhP is generally weak and practically absent in UV region, while the intensity of UV photoluminescence increases similar to the case of CL. It is also worth noting that the role of PhP is underestimated yet.

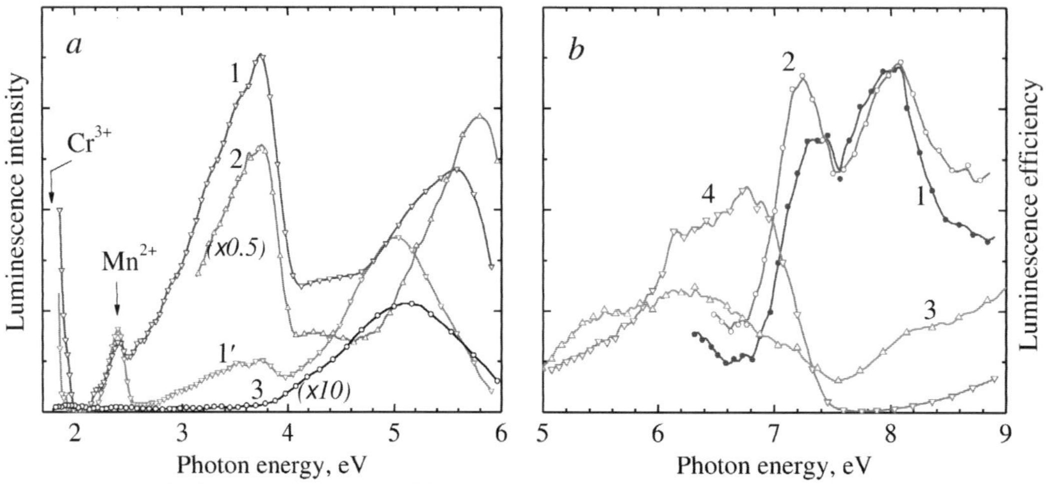

Figure 2. *a* – Emission spectra measured for two different 1:1 single crystals at the excitation by 8.1-eV photons at 6 K (curves 1 and 2) and RT (1'). Spectrum of phosphorescence measured at RT for the 1:1 crystal previously excited by 8.1-eV photons (curve 3, multiplied by a factor of 10). *b* – Excitation spectra for the emission measured at 5.3 eV (curve 1), 4.6 eV (2), 2.3 eV (3) and 1.8 eV (4) at 6 K.

Figure 2*b* presents the excitation spectra of different emissions measured for the 1:1 spinel crystal at 6 K. Based on these excitation spectra, two groups of emissions can be clearly separated. Firstly, impurity emissions of Mn^{2+} and Cr^{3+} (similar to CL spectra, peaks at 2.4 and ~1.8 eV, respectively) are efficiently excited strongly below the region of fundamental absorption (energy gap in $MgAl_2O_4$ at 6 K is expected around 8.1 eV). After emission spectra corrections, Cr^{3+} emission is too strong, dominates over other ones and is not shown in Figure 2*a*. Secondly, UV emissions above 3.5 eV have main excitation bands at ~7.2 eV and ~8 eV (excitation spectra for 4.6- and 5.3-eV emission are shown in figure) and can be considered as the intrinsic emissions of Mg-Al spinel crystals (see, e.g., [22]). The behavior of the 3.7-eV emission (not shown in Figure 2*b*) is more complicated, it is excited by photons of $h\nu < 7$ eV as well, but, in our opinion, can be related to ADs. Because of the overlapping of wide emission bands, the precise analysis of the origin of tentatively intrinsic emissions is rather complicated and should be continued. Nevertheless, the analysis of experimental PL data and similarity with $Y_3Al_5O_{12}$ and $Lu_3Al_5O_{12}$ single crystals ([25-28] and references therein) allow to suggest that the ~7.2 eV excitation band in our virgin spinel samples can be tentatively ascribed to ADs.

Figure 3*a* shows the fragments of TSL curves measured after irradiation of the 1:1 single crystal by 8-eV photons at 80 K or 270 K. Such photoexcitation leads to the appearance of the TSL peak at ~130 K and some other peaks at higher temperatures. Note that the low-temperature peak is located at higher temperature than that detected after electron irradiation at 6 K. Probably, higher temperature of photoexcitation as well as its low intensity causes such shift of the peak position. Similar to the case of crystal irradiation with an electron beam, the low-temperature TSL peak has a complex origin and cannot be approximated by the first order kinetics (extremely low values of frequency factor).

Figure 3*b* demonstrates the creation spectrum of the ~130 K TSL peak measured for the 1:1 crystal irradiated by a prescribed number of the exciting photons at each of several energies of 6.4–8.0 eV at 80 K. After every isodose irradiation, TSL was measured up to 450 K and the light sum of the 130 K peak (or peak intensity) was used for the construction of the "creation spectrum" reflecting the efficiency of the TSL peak induced by photons of different energies. The efficiency of TSL is high after the 1:1 crystal irradiation by photons of ~7–8 eV. Charge carriers needed for the further appearance of TSL could arise due to the formation of bound excitons localized, respectively, near negatively and positively charged $Mg|_{Al}$ and $Al|_{Mg}$ ADs as well as due to the beginning of band-to band transitions.

According to Figure 3*b*, this creation spectrum of the 130-K TSL peak is similar to the excitation (creation) spectrum of PhP for the same 1:1 sample (curve 2). In the latter case, the decay of PhP was recorded after an isodose irradiation by photons with a certain energy had been stopped. The PhP is detectable if the sample was previously excited by photons with the energy from 7 eV up to the edge of fundamental absorption around 8 eV (see transmission spectrum – curve 3). The measure of PhP was estimated via light sum of a signal, spectrally integrated or selected by an optical filter (3.2–5.1 eV) as well as the intensity measured after a certain time delay after the photoexcitation had been stopped. The shapes of PhP excitation/creation spectra coincide well for all three used registration methods.

Both creation spectra allow to elucidate the energy region of the exciting photons which cause a sequent appearance of conduction electrons/holes – either released from traps/structural defects or formed due to band-to band-transitions. It was mentioned already that ADs in $MgAl_2O_4$ are charged defects and serve as the traps for charge carriers. In particular, electrons can be released from $Al|_{Mg}$ with the corresponding optical absorption slightly above 7 eV (see also [12]). The similar situation has been earlier detected for ADs in $Lu_3Al_5O_{12}$ [26]. The sharp rise of PhP efficiency at $h\nu > 7.5$ eV can testify to the beginning of band-to-band transitions in the 1:1 spinel crystal at 270 K, while the efficiency decrease at $h\nu > 7.7–7.8$ eV reflects negligible penetration depth of such exciting photons (see transmittance spectrum in Figure 3*b*).

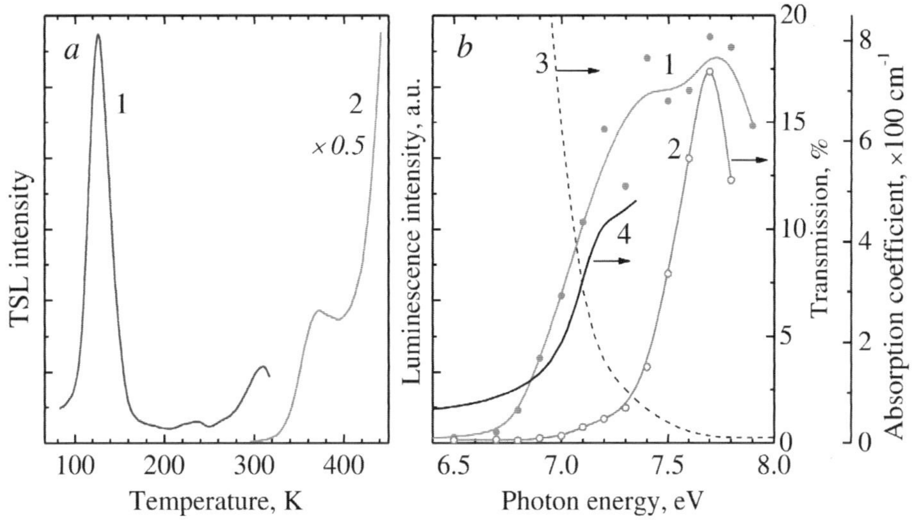

Figure 3. ***a*** – TSL curves measured for the 1:1 crystal irradiated by 8-eV photons at 80 K (curve 1) or 270 K (curve 2). ***b*** – Creation spectrum of the 130-K TSL peak by VUV radiation at 80 K (curve 1), excitation spectrum for phosphorescence (integral signal, curve 2) and transmission spectrum (3) for the 1:1 crystal at 270 K. Absorption spectrum of the 1:2.5 crystal (thickness of 0.1 mm) at RT (4).

Figure 3*b* also demonstrates the absorption spectrum measured in UV-VUV spectral region for an as-grown 1:2.5 crystal. Absorption coefficient in this spectral region has very high values, therefore the sample with thickness of 0.1 mm was used for this measurement. There is a clearly observed shoulder at 7.2 eV (less pronounced but still detectable in the 1:1 spinel), which was also detected in the excitation spectra of PL and tentatively ascribed to ADs (bound excitons near ADs, Figure 2*b*). Our previously obtained data on radiation-induced optical absorption in $MgAl_2O_4$ confirm the relation of the shoulder at ~7.2 eV to ADs being a constituent part of a number of the revealed paramagnetic defects, thermal annealing of which correlates with the attenuation of the 7.2-eV band of radiation-induced absorption [12]. As it was mentioned already, the concentration of ADs in the 1:2.5 nonstoichiometric crystal is significantly higher than that in the 1:1 crystal.

4. Conclusions

Manifestations of as-grown structural defects have been revealed in magnesium aluminate spinel single crystals with different stoichiometry by means of different optical methods. The main defects in virgin 1:1 and 1:2.5 spinel crystals are antisite defects – Mg^{2+} or Al^{3+} located in a "wrong" cation site,

$Mg|_{Al}$ or $Al|_{Mg}$. These ADs are charged with respect to the perfect lattice and serve as electron/hole traps or "stickers" for exciton-like electron excitations (bound excitons). Carrier traps manifest themselves via TSL arising at heating of the crystals irradiated with an electron beam or photons. Based on the excitation spectra of different emissions, optical absorption spectra as well as the "creation spectra" of TSL peaks by VUV radiation, the spectral region at ~7–7.5 eV, slightly below the energy gap, can be tentatively ascribed to ADs, the number of which is especially high in the 1:2.5 spinel single crystals. This suggestion is also confirmed by our previous study on radiation-induced defects (including ADs) in fast-neutron irradiated Mg-Al spinel crystals [12]. The excitation/creation spectrum of phosphorescence indicates on the beginning of interband transitions above 7.5 eV in $MgAl_2O_4$ at 270 K.

5. Acknowledgments

We are grateful to Drs E. Vasil'chenko and A. Maaroos for the help with experiments and useful discussions. This work has been carried out within the framework of the EUROfusion Consortium and has received funding from the Euratom research and training programme 2014-2018 under grant agreement No 633053. The views and opinions expressed herein do not necessarily reflect those of the European Commission. In addition, the research leading to these results has received funding from the Estonian Research Council – Institutional Research Funding IUT02-26.

6. References

[1] Ganesh I 2013 *Int. Mater. Rev.* **58** 63
[2] Clinard Jr F W, Hurley G F and Hobbs L W 1982 J. *Nucl. Mat.* **108-109** 655
[3] Sickafus K E, Minervini L, Grimes R W, Valdez JA, Ishimaru M, Li F, McClellan K J and T. Hartmann 2000 *Science* **289** 748
[4] Gonzales de Vicente S M, Hodgson E R and Shikama T 2017 *Nucl. Fusion* **57** 092009
[5] Gilbert A, Smith R, Kenny S D, Murphy S T, Grimes R W and Ball J A 2009 *J. Phys.: Condens. Matter* **21** 275406
[6] Dupree R, Lewis M H and Smith M E. 1986 *Phil. Mag. A* **53** L17
[7] Summers G P, White G S, Lee K H and Crawford J H 1980 *Phys. Rev. B* **21** 2578
[8] Bacorisen D, Smith R, Uberuaga B P and Sickafus K E 2006 *Phys. Rev. B* **74** 214105
[9] Zinkle S J 2012 *Nucl. Instrum. Meth. B* **286** 4
[10] Costantini J-M, Leiong G, Guillaumet M, Weber W J, Takaki S and Yasuda K 2016 *J. Phys.: Condens. Matter* **28** 325901
[11] Mironova-Ulmane N, Skvortsova V, Pavlenko A Feldbach E, Lushchik A, Lushchik Ch, Churmanov V, Ivanov D, Ivanov V and Aleksanyan E 2016 *Radiat. Meas.* **90** 122
[12] Lushchik A, Dolgov S, Feldbach E Pareja R, Popov A I, Shablonin E and Seeman V 2018 *Nucl. Instrum. Meth. B* **435** 31
[13] Ball J A, Pirzada M, Grimes R W, Zacate M O, Price D W and Uberuaga B P 2005 *J. Phys.: Condens. Matter* **17** 7621
[14] Nakonechnyi S, Karner T, Lushchik A, Lushchik Ch, Babin V, Feldbach E, Kudryavtseva I, Liblik P, Pung L and Vasil'chenko E 2006 *J. Phys: Condens. Mat.* **18** 379
[15] Feldbach E. Tõldsepp E, Kirm M, Lushchik A, Mizohata K and Räisänen J 2016 *Opt. Mater.* **55** 164
[16] Lushchik A, Feldbach E, Frorip A, Ibragimov K, Kuusmann I and Lushchik C 1994 J. *Phys.: Condens. Mat.* **6** 2357
[17] Lushchik A, Kudryavtseva I, Lushchik Ch, Vasil'chenko E, Kirm M and Martinson I 1995 *Phys. Rev. B* **52** 10069
[18] Skvortsova V, Mironova-Ulmane N and Ulmanis U 2002 *Nucl. Instrum. Meth. B* **191** 256
[19] Gritsyna V T, Afanasyev-Charkin I V, Kazarinov Yu G and Sickafus K E 2004 *Nucl. Instrum. Meth. B* **218** 264
[20] Gritsyna V T, Kazarinov Yu G, Kobyakov V A and Reimanis I E 2006 *Nucl. Instrum. Meth. B* **250** 342
[21] Sawai S and Uchino T 2012 *J. Appl. Phys.* **112** 103523

[22] Jozwik I, Jagielski J, Gawlik G, Jozwik P, Ratajczak R, Panczer G, Moncoffre N, Wajler A, Sidorowicz A and Thomé L 2016 *Phys. Chem. Minerals* **43** 439

[23] Lushchik A, Kirm M, Kotlov A, Liblik P, Lushchik Ch, Maaroos A, Nagirnyi V, Savikhina T and Zimmerer G 2003 J. Lumin. **102-103** 38

[24] Lushchik A, Lushchik Ch, Schwartz K, Vasil'chenko E, Kärner T, Kudryavtseva I, Isakhanyan V and Shugai A 2008 *Nucl. Instrum. Meth. B* **266** 2868

[25] Nikl M, Laguta V V and Vedda A 2008 *Phys. Status Solidi B* **245** 1701

[26] Babin V, Laguta V V, Makhov A, Nejezchleb K, Nikl M and Zazubovich S 2008 *IEEE Trans. Nucl. Sci.* **55** 1156

[27] Lushchik A, Kärner T, Lushchik Ch, Schwartz K, Savikhin F, Shablonin E, Shugai A and Vasil'chenko E 2012 *Nucl. Instrum. Meth. B* **286** 200

[28] Lushchik A, Lushchik Ch, Popov A I, Schwartz K, Shablonin E and Vasil'chenko E 2016 *Nucl. Instrum. Meth. B* **374** 90

IOP Conf. Series: Materials Science and Engineering **503** (2019) 012022 doi:10.1088/1757-899X/503/1/012022

Float zone single crystals for testing rods, pulled under electron beam heating

Anatoly Kravtsov[1], Kirils Surovovs[2] and Jānis Virbulis[2]

[1]KEPP EU AS, Carnikavas iela 5, Riga, Latvia
[2]University of Latvia, Department of Physics, Zellu iela 25, Riga, Latvia

E-mail: doc@keppeu.lv

Abstract. The article is devoted to the description of mathematical modelling and attempts to grow silicon single crystals from a pedestal. The crystals are intended to be used for impurity composition testing in rods with a diameter of 300 mm grown with electron beam heating. The testing is being planned both by the method of FTIR spectroscopy and by functional testing of devices that might be manufactured using single crystals grown from pedestal. The article also describes the improvements of equipment, which were necessary for crystal growth attempts, and substantial difficulties that occurred in the process and hindered single crystal growth, allowing to obtain only a polycrystalline sample.

1. Introduction

KEPP EU developed new method of achieving polycrystalline silicon rods by electron beam heating method (EB rods), which are intended for future applications like float-zone (FZ) single crystal growth. The developed method allowed to obtain rods of 300 mm in diameter [1], which is remarkable, because electromagnetically heated FZ technology for 300 mm single crystals has not been developed yet.

Main requirements to such rods are purity, stable prescribed geometry and mechanical strength. The content of main impurities, such as boron and phosphorus, must be less than 10 ppta. Detection of such low level of contaminants in silicon is possible only by FTIR Spectroscopy at liquid He temperature. For these tests, EB rods should be melted and recrystallized as single crystals. To reduce additional melt contamination during this stage, crucible-free methods should be used. The present article is focused on a pedestal method due to its relative simplicity in comparison with FZ method.

In the pedestal method, electromagnetic (EM) inductor is used to melt upper surface of the polycrystalline rod, which is called pedestal, and single crystal is pulled upwards from the molten zone. Additional pedestal heating from the side may be used to increase the deflection of the melting interface and thus reduce the chance of melt freezing.

The work consists of three parts: modelling of crystal growth from pedestal, creation of a puller based on available laboratory equipment and the actual growth of single crystals. The modelling processes were carried out for the growth of control rods of 12-20 mm in diameter [2].

2. Modelling

The numerical model was quasi-stationary and included only cylindrical phase of crystal pulling. To reduce time consumption and allow for numerous calculation studies, axially symmetrical approximation was used, and the melt flow was not considered. Inductor was optimized to increase melt height at the system axis and therefore decrease the risk of melt freezing. More details about the calculation procedure are described in [3].

Figure 1 shows the temperature distribution, calculated for different diameters of crystal and pedestal. The obtained optimal values of integral electromagnetic heat Q_{EM} and additional pedestal side heating Q_{add} are disproportionally increasing for larger pedestals, raising the ratio Q_{EM}/Q_{add}. The position of additional heater was optimized using heat transfer calculations with FEMM program in simplified axially symmetrical representation of a flat pedestal, see Figure 1, right top. Heat fluxes $q = \lambda \frac{\partial T}{\partial n}$ were imposed as boundary conditions on all pedestal surfaces except bottom surface, where fixed temperature was used to ensure calculation convergence. Here λ is heat conductivity and n – normal coordinate. Calculations predicted that the optimal position of heater, which creates maximal temperature difference ΔT between the pedestal centre and its rim, is dependent on pedestal radius. Thus optimal z_Q (the distance from the additional heater to pedestal edge) increases from 30 mm for 37.5 mm pedestal to 40 mm for 50 mm pedestal. Moreover, ΔT becomes less sensitive on z_Q when pedestal radius increases.

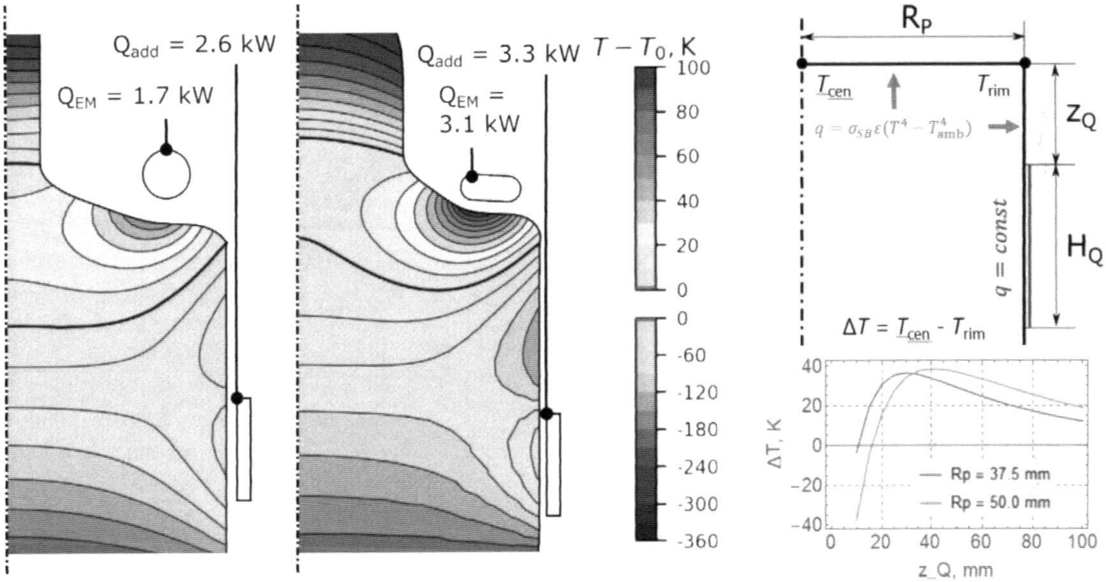

Figure 1. Left: temperature field in silicon (T_0 = 1687 K is melting temperature), for the system with 12 mm crystal grown from 75 mm pedestal. Middle: temperature field for the system with 44 mm crystal grown from 100 mm pedestal. Integral values of electromagnetic heating Q_{EM} (black oval corresponds to the inductor) and additional pedestal side heating Q_{add} (rectangle corresponds to the additional heater) are given. Right top: the scheme of simplified geometry and boundary conditions for additional heater position optimisation. Right bottom: temperature difference between the pedestal centre and its rim, over additional heater position, calculated for different pedestal radii R_p.

3. Process development

3.1. Puller creation

Vacuum furnace for laboratory electron beam melting (Figure 2, left) was chosen as a base machine for processes with induction heating. A standard high-frequency generator FTEC 30 (Figure 2, middle) was used as a power source.

The single-circuit design of the FTEC 30 was replaced by a two-circuit design, which is energetically more efficient [4]. In order to increase the efficiency of the system and reduce the influence of uncontrolled induced currents, we have changed constructively the HF tube, primary (anode) circuit, the circuit of separating capacitors, and inter-loop air transformer. Then we have added a tank of capacitors and have put all of that in a separate unit (Figure 2, right).

The power required for crystal growth processes was obtained from the results of numerical modelling and based on the literature data [5]. According to the model (see Figure 1), an upper estimate of total power required for 44 mm crystal growth is approximately 7 kW. On the other hand, the comparison of the power consumed by power suppliers with the capacity to grow single crystals in Steremat FZ1510 and Leybold-Hereus 1450/35 shows that the efficiency of such systems is 5-8%. It suggests that if the main coil power consumption (for single crystal diameter up to 44 mm) is around 3 kW, the power of the FTEC 30 generator is sufficient for the process.

Figure 2. Vacuum furnace for electron beam melting (left), FTEC 30 HF power supply system (middle) and HF part of upgraded power source (right) where 1 designates capacitor tank, 2 – inter-circuit transformer, 3 – capacitors of the primary (anode) circuit, 4 – capacitors of the divide (separating) circuit, and 5 – HF tube.

3.2. Growth experiments

The rod with diameter of 75 mm, pulled under electron beam heating, was used as a pedestal. It had been obtained using the equipment described in [1]. The purity of silicon had not been measured, as the rod was intended only to be used in pedestal process development. However, the potential purity of EB rods was shown to be comparable with that of the high-purity reference sample of FZ silicon [1].

Despite the long work to harmonize the HF tube with the electrical load, it was possible to melt the pedestal upper surface only partially (see Figure 3, left). The power supplied by the bottom heater was changed from 2.6 kW (predicted by the model) to the value that was sufficient to melt pedestal side surface (see figure 3, right). However, it did not help to fully melt the upper surface of the pedestal.

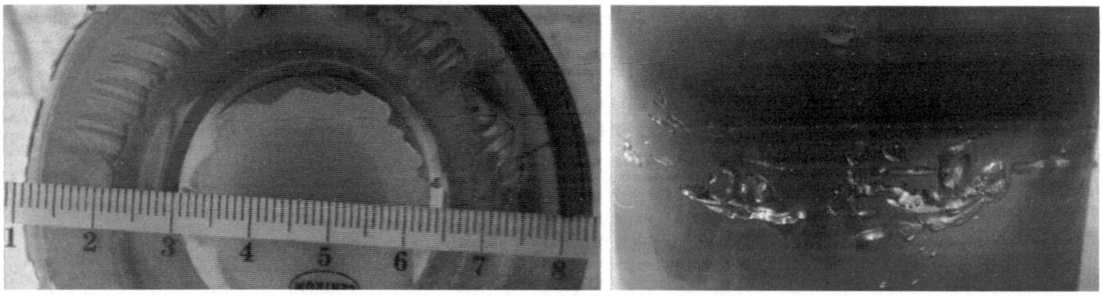

Figure 3. Pedestal upper surface, melted by EM heating (left) and side surface, melted by additional heating (right).

The power supply was increased by replacing the transformer with a more powerful (+ 20%), regular HF tube FU 307S at ITK 30-2 (+ 300%). We have appropriately matched the HF tube with the electrical load, and it allowed to melt the upper surface of pedestal completely (Figure 4, left). We successfully connected the seed crystal to the melt (Figure 4, right). However, it was not possible to pull a single crystal due to significant risk of spontaneous crystallization.

The described result was obtained using a coil that was manufactured according to the model [2]. After trying to conduct a process using a coil with the prolonged bottom part to concentrate EM field closer to the pedestal centre (similar to the one used in [6]) we have not obtained a positive result. The photographs of both coils and the pedestal with a non-molten centre are shown in Figure 5, left.

These difficulties were overcomed by tuning of system parameters and improving heat flux distribution in the system. The first growth process was performed and irregular-shaped polycrystalline rod with diameter of 11-16 mm and length of 159 mm was obtained (see Figure 5, right). The shape of it demonstrates that the growth process is highly unstable, and its control is very challenging.

An important discrepancy between the mathematical model and the experiment should be addressed. The former demonstrated the existence of a steady-state solution for different inductor parameters and crystal diameters, thus predicting good stability of the process, at least for relatively small crystals. The experimental growth process, however, was highly unstable. Some of the possible explanations are *T* gradient oscillations induced by non-stationary melt flow, 3D features of one-turn EM inductor, insufficiently sensitive power control in the experiment and different ambient conditions for thermal radiation and EM field. Although some of the mentioned effects, such as melt flow, were investigated previously [2], additional numerical studies are needed to improve the validity of the mathematical model by detecting which simplifications induce discrepancies between the model and the experiment.

Figure 4. Pedestal upper surface during the process and after (left) and seed crystal during the process and after (right).

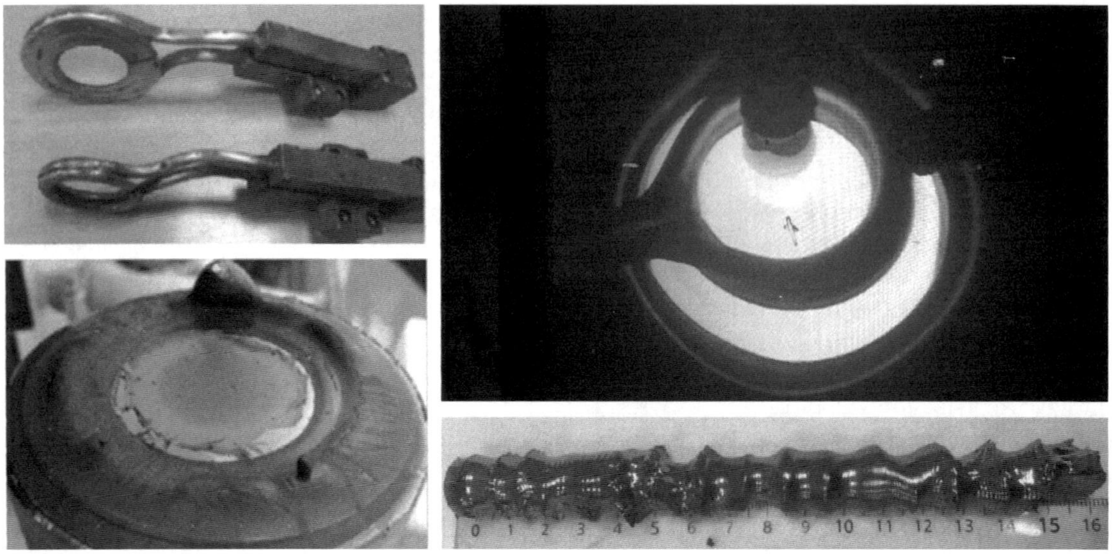

Figure 5. Top left: photograph of the used coils: the coil manufactured in accordance with the model (lower coil) and the coil with the modified middle part, which is closer to pedestal centre (upper coil). Bottom left: fused pedestal with a non-molten centre. Top right: the growth process with improved thermal conditions. Bottom right: irregular-shaped polycrystalline sample that was obtained from the most successful growth process.

4. Conclusions

It was shown that the laboratory furnace for electron beam melting can be used as a base for creating equipment for pedestal method of single crystal growth. The inductor, proposed by numerical modelling, was capable to create the molten zone and the simulated necessary power of bottom heater was close to the experimental one.

The process with pedestal diameter 75 mm and seed crystal with diameter 12 mm was performed and polycrystalline sample was pulled. Its irregular shape demonstrates that the process is unstable, therefore the used quasi-stationary model might be insufficient for process development, and time-dependent models could be more useful.

Acknowledgments

Research is supported by the Competence Center MASOC project ID no. 1.2.1.1/16/A/003, investigation 2.7. "Investigation of technology growth of silicon single crystals by FZ pedestal technique"

References

[1] Kravtsov A 2018 *IOP Conf. Ser.: Mater. Sci. Eng.* **355** 012007
[2] Surovovs K, Plāte M and Virbulis J 2017 *VIII Internat. Sci. Colloq. Mod. for Mater. Proc.* p. 37-42
[3] Surovovs K, Kravtsov A, Sabanskis A and Virbulis J 2018 *6th Eur. Conf. on Cryst. Growth* S02–P34
[4] Vasiljev A 1979 *Lamp generator for HF heating* (Leningrad: Mashinostrojenie)
[5] Ratnikov D 1976 *Crucible free melting* (Moscow: Metallurgija)
[6] Wünscher M, Lüdge A and Rieman H 2011 J. Cryst. Growth **318** 1039

Influences of sputtering process factors on wear resistance of TiN coated on a machine component

Charnnarong Saikaew

Department of Industrial Engineering, Faculty of Engineering, Khon Kaen University, Khon Kaen 40002 Thailand

E-mail: charn_sa@kku.ac.th

Abstract. In this work, effects of sputtering process factors including DC current, pressure and Ar to N_2 flow rate ratio on wear resistance of TiN coated on cast stainless steel of a fishing net-weaving machine part namely, an upper hook were systematically investigated using 2^k factorial design, analysis of variance (ANOVA) and other statistical tools. Wear testing experiments of the TiN coated upper hooks were carried out on the real fishing net weaving machine. A normal probability of effects and main and interaction effect plots of process factors were constructed in order to identify the statistically significant process factors and to determine an appropriate operating condition of the process factors with the lowest weight loss. The Ar to N_2 flow rate ratio and the interaction between the pressure and the Ar to N_2 flow rate ratio were found to be statistically significant while pressure and sputtering DC current were not statistically important.

1. Introduction

Titanium nitride (TiN) is widely used as a coating to harden and protect cutting and sliding surfaces in abundant tribological applications [1]. It has a cubic structure of NaCl type with modulus of elasticity of 250-450 GPa and high Vickers hardness of 18-21 GPa [2]. Its surface and material properties are considerably dependent on its creation process. The most common techniques for TiN coating are physical vapor depositions (PVD) including sputtering, filtered cathodic arc and electron beam evaporation and chemical vapor deposition (CVD) [3]. Sputtering is the most generally used method because of high quality thin film coating and well-controlled process. Normally, TiN is produced by reactive sputtering, in which sputtered Ti atoms are reacted with nitrogen ions to form TiN molecules [4].

Oblique angle deposition (OAD) technique is a modified deposition method, in which substrate is tilted at an angle greater than 70° with respect to the normal of deposition direction and rotated at a suitable speed [5]. It provides a functional control of surface nanostructures by shadowing effect and surface diffusion [6]. The main advantage of this technique is ability to control the diameter, shape density of nanostructures by varying the deposition conditions including rotation speed, oblique angle, operating pressure, gas mixing ratio and sputtering power. The technique is applied to TiN hard coating to produce high-density nanocolumnar structure, which could have an extreme hardness, low friction coefficient and very good wear resistance [7].

In this work, TiN was coated on cast stainless steel of a fishing net-weaving machine component namely, an upper hook by OAD technique with DC magnetron sputtering. A three factor, two level full factorial design was selected to investigate the effects of the three sputtering process factors on average weight loss and determine an appropriate level of each process factor with minimization of weight loss.

2. Materials and procedure

The cast stainless steel upper hooks were made by the lost wax casting process and were machined by turning and milling operations. Chemical composition of the workpiece material consisted of 62.28% Fe, 21.99% Cr, 6.91% Ni, 3.12% Cu, 2.2% Si, 1.94% Mn and 1.56% Al. A vibratory finishing machine with mixed ball burnishing media in different sizes ranging from 5 to 8 mm was used for surface finishing. The oblique angle coating system included a high vacuum chamber equipped with a 3″ target magnetron gun, 600 W radio frequency generator, 400 W DC power supply and a turbomolecular pump. Figure 1(a) illustrates an oblique angle sputtering schematic diagram. The titanium nitride was deposited on the upper hooks by reactive sputtering of pure titanium target under a mixture of argon (Ar) and nitrogen (N_2) gases. The oblique angle and rotation speed were fixed at 70° and 45 rpm, respectively. Figure 1(b) depicts a typical photograph of the upper hooks coated by TiN. Three process factors including DC current, operating pressure and Ar to N_2 flow rate ratio were simultaneously investigated using the full factorial design method. The ranges of sputtering DC current, pressure and Ar to N_2 flow rate ratio were 0.35 to 0.45 A, 50 to 100 Pa and 0.5 to 1.5, respectively. The TiN coatings were systematically performed according to the 2^3 factorial design as shown in Table 1. The TiN coated upper hooks were then used to carry out the wear testing experiments on the fishing net-weaving machine with direct sliding between the upper hooks and the fishing net (nylon 6). After carrying out the wear testing experiments, each upper hook was cleaned with a dry cloth and then wiped clean with alcohol. Each hook was weighed with 4-digit scale and the weight loss data was recorded. The analysis of variance (ANOVA) was employed to investigate the influences of the three sputtering process factors on the wear resistance according to designed experiments.

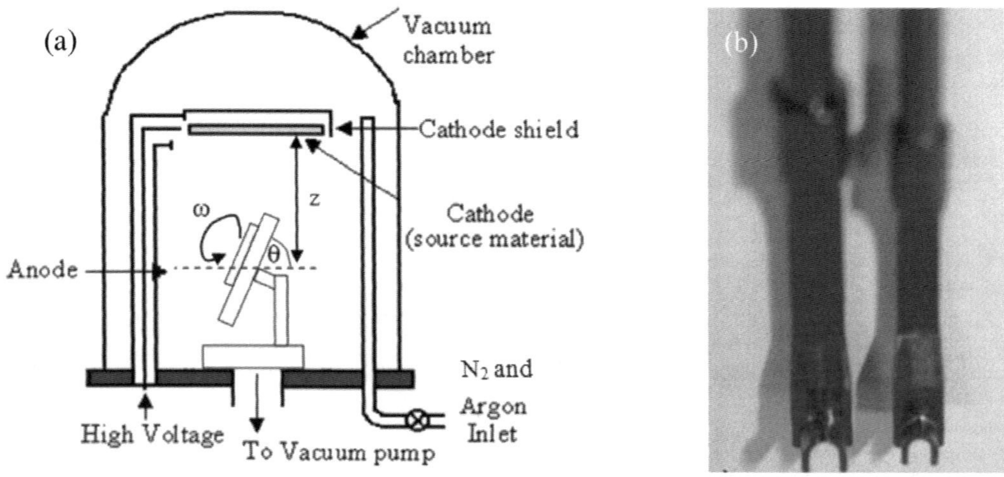

Figure 1. (a) An oblique angle sputtering schematic diagram (b) a typical photograph of upper hooks coated by TiN

Table 1. Factors and their levels in TiN coated on cast stainless steel process

Experimental run	DC current (A)	Pressure (Pa)	Ar /N_2
1	0.35 (-)	50 (-)	0.50 (-)
2	0.45 (+)	50 (-)	0.50 (-)
3	0.35(-)	100 (+)	0.50 (-)
4	0.45 (+)	100 (+)	0.50 (-)
5	0.35 (-)	50 (-)	1.50 (+)
6	0.45 (+)	50 (-)	1.50 (+)

7	0.35 (-)	100 (+)	1.50 (+)
8	0.45 (+)	100 (+)	1.50 (+)

3. Results and discussion

Before performing ANOVA, the data set of weight loss values should be evaluated whether it distributes as normality. The probability plot was used to illustrate the data points, fitted line of the data points and associated confidence intervals (C.I.) based on parameters estimated from the data set along with an Anderson-Darling (AD) goodness-of-fit statistic and associated *p*-value [8]. Figure 2(a) illustares probability plot of weight loss value. The probability plot shows that the plotted points of the complete data set of weight loss roughly form a straight line and fall within the 95% confidence interval. Moreover, AD statistic of the data set was small with high *p*-value compared to the level of significance of 0.05. This confirmed that normal probability distribution fitted the data set of the weight loss values capably. Furthermore, the adequacy of the underlying model should be checked using a primary diagnostic tool, residual analysis. Figure 2(b) displays a normal probability plot of residuals. If the underlying error distribution is normal, this plot will resemble a straight line [8]. Obviously, the normal probability plot of residuals for weight loss did not indicate anything particularly troublesome problem of normality. Hence the residual analysis was satisfactory.

The results from the 2^3 full factorial experiment were analyzed by a half normal probability plot of effects using the Design Expert software package [9]. A half normal probability plot of effects is a plot of the absolute value of the effect estimates against their cumulative normal probabilities [8]. The significant effects with nonzero means will not lie along the straight line while the effects that are imperceptible are normaly distributed, with mean zero and variance σ^2 and will contribute to fall along a straight line on this plot. The straight line on the half normal plot always passes through the origin and passes close to the fiftieth percentile data value. Figure 3(a) presents the half normal plot of the effects for weight loss value. The important effects for the weight loss were factor *C* (Ar to N_2 flow rate ratio) and interaction between factor *B* (pressure) and factor *C*. Factor *B* was included to preserve hieraarchy in the model. This implied that if the *BC* interaction was in the model, both the main effects *B* and *C* should be included. Pareto chart of effects is another tool used to evaluate the effects of the process factors. Figure 3(b) illustrates the Pareto chart of effects for this study. The lower horizon line indicates Student's *t*-test value for the minimum statistically significant effect magnitude for a level of significance of 0.05. Only factor *B* and *BC* interaction were higher than the threshold. Like the half normal plot of the effects, factor *B* was included to preserve hieraarchy in the model. Thus, only pressure, Ar to N_2 flow rate ratio and the interaction between pressure and Ar to N_2 flow rate ratio were statistically significant at the level of significance of 0.05.

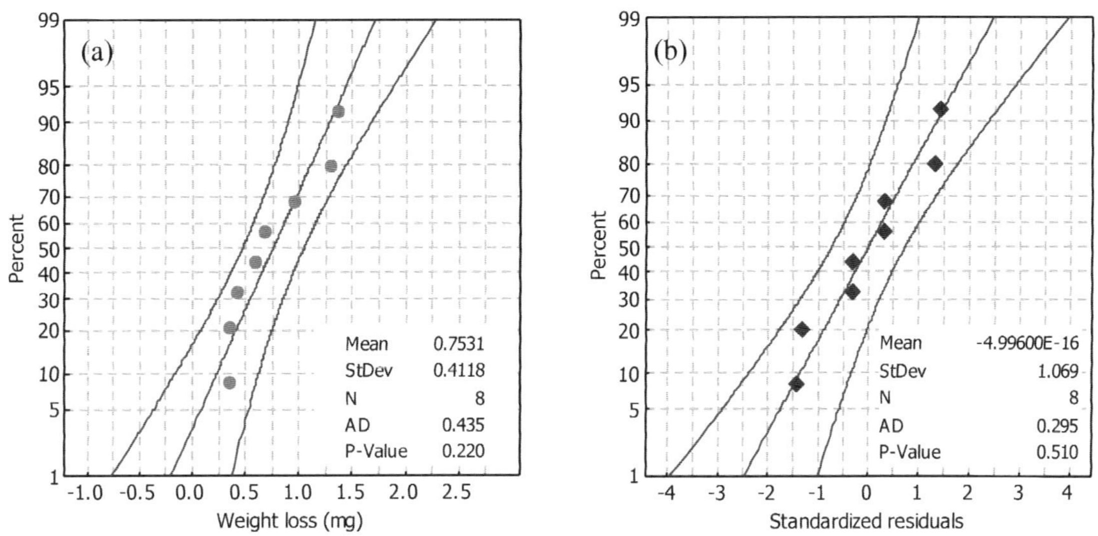

Figure 2. (a) Probability plot of weight loss value and (b) normal probability residual plot for weight loss

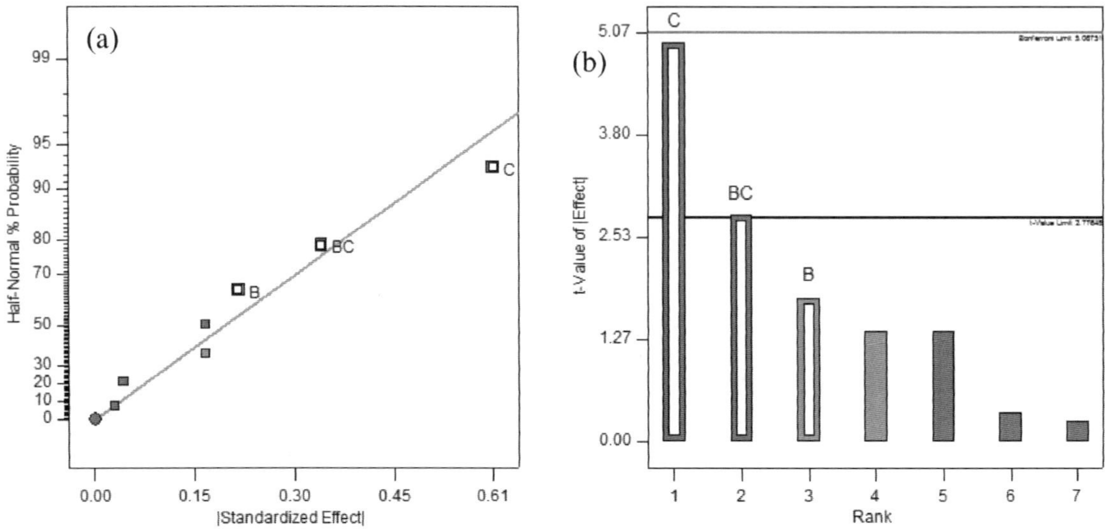

Figure 3. (a) Half normal probability plot of effects and (b) Pareto chart of effects for weight loss value

Table 2 shows the effect estimates, sum of squares and percent contribution of the three sputtering factors on the weight loss of the TiN coated upper hooks. Based on variance component estimates, 61.94% of the total variation in weight loss value was due to the difference between the Ar to N_2 flow rate ratio levels while 19.91% resulted from the differences between pressure levels and Ar to N_2 flow rate ratio levels. Hence Ar to N_2 flow rate ratio was approximately three times as influencial as the interaction between the pressure and the Ar to N_2 flow rate ratio.

Table 2. Effect estimates, sum of squares and percent contribution

Model term	Effect estimate	Sum of squares	Percent contribution
A-DC current	0.17	0.057	4.80
B-Pressure	0.22	0.096	8.06
C-Ar/N_2	-0.61	0.740	61.94
AB	-0.17	0.057	4.80
AC	-0.044	0.004	0.32
BC	-0.34	0.240	19.91
ABC	-0.031	0.002	0.16

The effects of the statistical significance of the three sputtering factors on the weight loss of the TiN coated upper hooks were evaluted by ANOVA as shown in Table 3. The results in Table 3 were analyzed with ANOVA used for investigating the influence of the three sputtering factors on response weight loss. Table 3 reveals that the model F-value of 11.89 with a p-value of 0.0184 was adequate. A coefficient of determination (R^2) of 0.8992 defined as the ratio of the explained variation to the total variation and was a measure of the degree of fit. The adjusted R^2 of 0.8235 was in reasonable agreement. This concluded that the Ar to N_2 flow rate ratio and the interaction between the pressure and Ar to N_2 flow rate ratio significantly affected the mean weight loss at the level of significance of 0.05.

The statistical model of weight loss as the function of pressure, Ar to N_2 flow rate ratio and the interaction between the pressure and the Ar to N_2 flow rate ratio was expressed in Eq. **(1):**

$$weight\,loss = 0.75 + 0.11B - 0.3C - 0.17BC \quad (1)$$

where *weight loss* represented the response variable, *B* denoted pressure and *C* was Ar to N_2 flow rate ratio. The equation was in terms of coded factors on the -1, +1 scale.

Table 3. ANOVA table for weight loss

Source	Df	SS	MS	F-value	p-value
Model	3	1.07	0.36	11.89	0.0184
B-Pressure	1	0.09	0.09	3.20	0.1482
C-Ar/N_2	1	0.74	0.74	24.57	0.0077
BC	1	0.24	0.24	7.90	0.0483
Error	4	0.12	0.03		
Total	7	1.19			
	PRESS = 0.48	$R^2 = 0.8992$	Adj. $R^2 = 0.8235$		

Figure 4(a) presents a plot of actual weight loss versus the predicted weight loss values. This plot shows the prediction capability of the model. The relationship between the actual value and the predicted value of weight loss was expressed in Eq. (2):

$$\hat{P} = 0.076 + 0.8992\,weight\,loss \quad (2)$$

where $\hat{P}$ represented the predicted weight loss whereas *weight loss* denoted the actual value. The R^2 was 0.899 indicating the model in Eq. (2) accounted for 89.9% of the variability in the data. The intercept of the linear model was close to zero and the slope was about 1. The plot indicated that model could satisfactorily be used in predicting the weight loss response. Similarly, Figure 4(b) also shows the pair difference between the actual value and the predicted value of weight loss in each experimental run. This plot clearly indicated that there was a small difference between the actual value and the predicted value of weight loss in each experimental run. This also confirmed that the model in Eq. (1) was adequate.

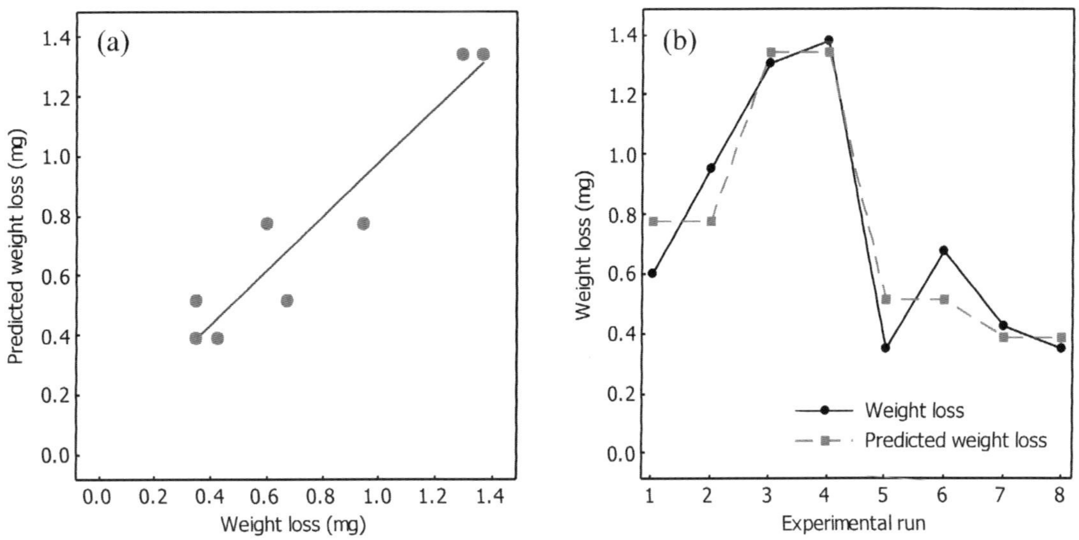

Figure 4. (a) Scatter plot between actual weight loss and predicted weight loss and (b) the pair difference between the actual value and the predicted value in each experimental run

The objective of this study was to minimize weight loss of the TiN coated upper hooks after carrying out the wear testing experiments. To assist in accomplishing this objective, graphical methods can be effectively used. A main effect plot is a graph of the averages at each level of a factor [8]. A main effect occurs when the average response changes across the levels of a factor. The main effect plot is employed to compare the relative strength of the effect of the factor. Although this plot is useful to compare main effect, the ANOVA table is a primary tool used to significantly assess the influence of the factor. The main effect plots displayed in Figures 5(a) and 5(b) prove that there were no statistically significant differences of the averages weight loss values between the two levels of the factor DC current and the factor pressure. On the other hand, there was a statistically significant difference between the two levels of the factor Ar to N_2 flow rate ratio as shown in Figure 5(c). The main effects do not have much meaning when they are involved in statistically significant interactions [8]. Hence it is necessary to evaluate any interractions that are important. In order to interpret interaction effectively, an interaction plot was constructed as illustrated in Figure 5(d), (Pressure and Ar/N_2 interaction). The non-parallel lines revealed that there was a strong interaction between pressure and Ar to N_2 flow rate ratio. The interaction indicated that Ar to N_2 flow rate ratio had little effect at high (+) Ar to N_2 flow rate ratio but a high positive effect at low (-) Ar to N_2 flow rate ratio. The pressure effect was very small when the Ar to N_2 flow rate ratio was at the high level and very large when the Ar to N_2 flow rate ratio was at the low level. However, the effect of Ar to N_2 flow rate ratio was not different at low level of pressure. Therefore, the minimum weight loss value was obtained when pressure and Ar to N_2 flow rate ratio were at low levels (i.e., pressure at 50 Pa and Ar to N_2 flow rate ratio at 0.5). This would allow the reduction of the Ar to a lower production cost. The previous studies have found that as a flow rate of Ar decreases, the grain size of TiN coating decreases resulting higher hardness as well as wear resistance improvement [1,10,11]. In addition, the effect of DC current was not different. Hence weight loss value was prefered for lower level of DC current.

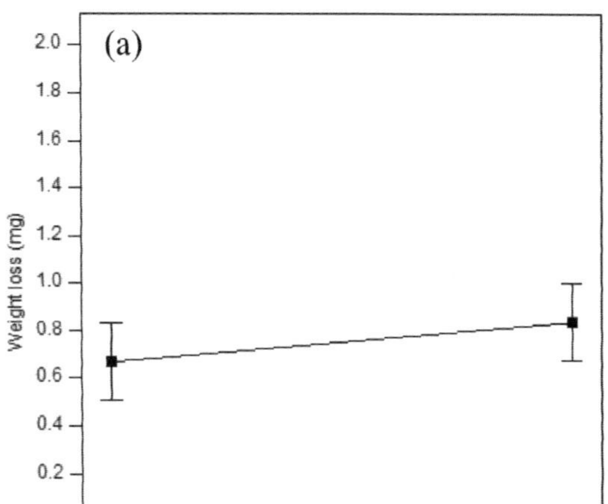

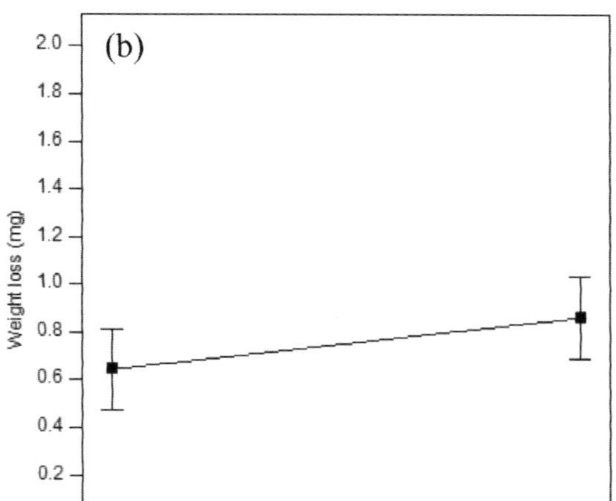

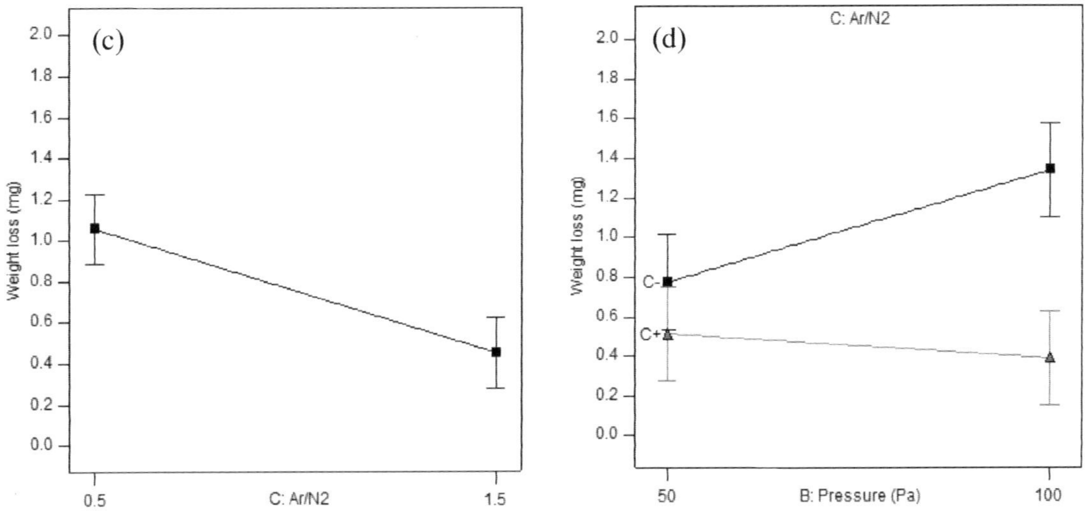

Figure 5. Effect graphs for factors (a) DC current (b) pressure (c) Ar to N_2 flow rate ratio and (d) interaction between pressure and Ar to N_2 flow rate ratio

4. Conclusion

This study had shown how statistically designed experiments combined with other statistical techniques could be used for investigating the influences of sputtering process factors affecting weight loss of TiN coated fishing net weaving machine component during carrying out the wear testing experiments. A normal probability of main and interaction effects of process factors was plotted against cumulative probability in order to identify the statistically significant process factors. The Ar to N_2 flow rate ratio and the interaction between the pressure and the Ar to N_2 flow rate ratio were found to be statistically significant at the level of significance of 0.05. An appropriate operating condition of the process factors was at DC current of 0.35 A, pressure of 50 Pa and Ar to N_2 flow rate ratio of 0.5.

Acknowledgements

This research is financially supported by the Thailand Research Fund (TRF) under the Grant no. RSA 5980025.

References

[1] Bahri A, Guermazi N, Elleuch K and Urgen M 2015 *Wear* **342-343** 77-84
[2] Stone D S, Yoder K B, and Sproul W D 1991 *J. Vac. Sci. Technol. A*,**9(4)** 2543–7
[3] Kim G S, Lee S Y, Hahn J H, Lee B Y, Han J G, Lee J H and Lee S Y. 2003 *Surf. Coat. Technol.* **171** 83–90
[4] Pham V H, Yook S W, Lee E J, Li Y, Jeon G, Lee J J, Kim H E and Koh Y H 2011 *J. Mater. Sci.* **22** 2231-7
[5] Knorr T G and Hoffman R W 1959 *Phys. Rev.* **113** 1039-46
[6] Lintymer J, Martin N, Chappe J M, Delobelle P, Takadoum J 2004 *Surf. Coat. Technol.* **180C–181C** 26–32
[7] Lintymer J, Martin N, Chappe J M and Takadoum J 2008. *Wear* **264** 444-9.
[8] Montgomery D C 2001 *Design and Analysis of Experiments* 6th ed John Wiley & Sons New York 2001
[9] Stat Ease, Inc. Design Expert Version 7 User's Manual. Stat Ease, Inc. Minneapolis, MN. (2005)
[10] Raoufi M, Mirdamadi S and Mahboubi F 2011 *Surf. Coat. Technol.* **205** 4980-4
[11] Nishat A, Junqing L, Yun K J, Chan G L, Jae H Y and Faheem A 2012 *Mater. Chem. Phys.* **134** 839-44

IOP Conf. Series: Materials Science and Engineering **503** (2019) 012024 doi:10.1088/1757-899X/503/1/012024

Features of nanostructure and functional properties formation in Ti-Ni shape memory alloys subjected to quasi-continuous equal channel angular pressing

I Yu Khmelevskaya[1,*], R D Karelin[1], S D Prokoshkin[1], M G Isaenkova[2], Yu A Perlovich[2], V A Fesenko[2], V S Komarov[1] and M M Zaripova[2]

[1]National University of Science and Technology MISiS, Russia
[2]National Research Nuclear University MEPhI, Russia

[*]khmel@tmo.misis.ru

Abstract. Influence of various deformation regimes, included severe plastic deformation by multi-pass rolling, rotary forging and equal channel angular pressing (ECAP) in the normal and quasi-continuous modes, on the structure and properties of Ti-Ni shape memory alloys was studied and compared. Features of structure formation providing a combination of high functional properties were analyzed. Investigation of the texture development, depending on the ECAP regimes and post-deformation annealing, was made.

1. Introduction

Ti-Ni-based shape memory alloys (SMA) are advanced functional materials. They allow implementing the performance characteristics of devices and technologies that are not achievable by using other materials [1-3]. Severe plastic deformation (SPD) is one of the most attractive methods of thermomechanical treatment that allows obtaining ultrafine-grained (UFG) structure and considerably improving mechanical and functional properties of Ti-Ni SMA without changes in chemical composition [4-8]. UFG structure is generally ranged into nanocrystalline (grain/subgrain size is less than 100 nm) and submicrocrystalline (grain/subgrain size from 100 nm to 1 μm) structures [9]. In accordance with the previous experience the best combination of mechanical and functional properties in Ti-Ni shape memory alloys can be achieved after formation of a completely nanocrystalline structure, with the size of structural elements 40-80 nm [10]. This structure was obtained in thin samples with a thickness of about 0.1 mm after multi-pass cold rolling. In order to obtain bulk nanostructured billets from Ti-Ni SMA the most promising SPD processes such as equal channel angular pressing (ECAP) and rotary forging at low temperatures were used [11, 12]. It was shown in previous studies that six-eight passes of traditional ECAP with the channels intersection angle of 110° and at the temperature not higher than 450 °C provides formation of mainly equiaxed ultrafine-grained (submicrocrystalline) structure. For the further grain refinement the number of passes was increased, and the deformation temperature was decreased. However, even after ECAP at 350 °C, the grain size remained in the submicrometer range of sizes (100-200 nm) [12]. Thus, for the very last time, the ECAP method allowed obtaining only a mixed submicrocrystalline structure. In this context, quasi-continuous ECAP was performed with the channel intersection angle of 120° at a temperature of 400 °C in order to obtain completely nanocrystalline structure in bulk samples [13]. The main features of the structure and functional properties in combination with the analyzes of the texture, obtained after ECAP and rotary forging with post-deformation annealing (PDA), in comparison with the multi-pass cold rolling, are described and discussed in the present work.

2. Experimental procedure

In the present work, TN-1 (from Ti-50.0 to Ti-50.2 at.% Ni) alloy, supplied by "Industrial Center MATEK-SMA Ltd.", was studied. The billets for rotary forging and ECAP, rods 20 mm in diameter, were produced by screw rolling at 850-950 °C with reduction of 7-20 % per pass and interpass heating (hot-rolled state). ECAP was carried out in the traditional regime with reheating between passes at 450 °C for 20 passes and in the quasi-continuous regime at the temperature of 400 °C for 3, 5, and 7 passes. The channels intersection angle for traditional and quasi-continuous modes was 120°. The rotary forging was made from a diameter of 20 to 5 mm at 450 and 350 °C with relative strain from 1 to 5 % per one pass. The structure was studied at room temperature using "JEM-2100" transmission electron microscope. To analyze the crystallographic texture of deformed samples by direct pole figures (DPF) automated diffractometer "DRON-3", equipped with texture set, was used. This method consists in the consequential recording of the integrated intensity of the selected X-ray reflection from the {hkl} planes with rotation and inclination of the sample around the normal to the selected sample section [14]. The reflection intensity {hkl}, recorded for different sample orientations, was used to obtain a partial textural direct pole figures DPFs {100}, {110} and {112} for section perpendicular to the axis of the studied rod. The main axes of deformation are the compression direction (CD) and the tension direction (TD) for the last pass. For investigation of B2-austenite texture, the sample was heated to a temperature of 80-120 °C. A texture comparison was performed after ECAP and after rotary forging. The mechanical properties were determined at room temperature by the uniaxial tensile tests using universal tensile machine "INSTRON 3382" with a deformation rate of 4 mm/min. The maximum completely recoverable strain was estimated by a thermomechanical method using band samples.

3. Results and discussion

3.1 Microstructure

The structure of flat samples with a thickness of 0.13 mm obtained by multi-pass cold rolling and post-deformation annealing (PDA) by various regimes is shown in Figure 1. Directly after multi-pass cold rolling (accumulated true strain e = 1.9) a mixed nanocrystalline and amorphous structure was formed (Figure 1). During the PDA, the amorphous structure crystallizes into a nanocrystalline, the new grains (obtained both due to deformation and crystallization of the amorphous structure) grown, but their average size is less than 100 nm up to 400 °C. The main features of this structure are, firstly, its "bimodal" character: some grains retain a size of 5-20 nm, while others grow to a size of 30-80 nm and, secondly, a low dislocation density after crystallization from the amorphous phase.

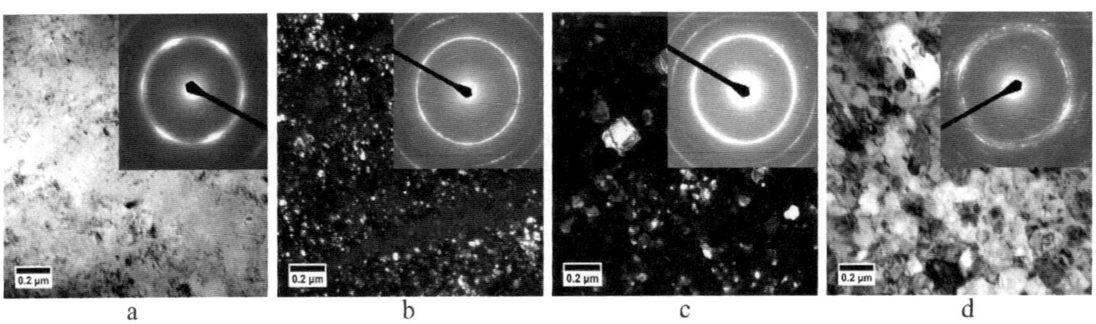

Figure 1. Microstructure of the TN-1 alloy subjected to multi-pass cold rolling (MPCR) according to various regimes: (a) MPCR, (b) MPCR + 250 °C, (c) MPCR + 350 °C and (d) MPCR + 400 °C. Transmission electron microscopy: dark-field images, electron-diffraction patterns in the insets.

During the process of ECAP, principally different mechanism of the ultrafine-grained structure formation as compared with cold rolling appears, comprising dynamic polygonization and recrystallization. Due to the repeated changes in the deformation direction, band structures (deformation localization bands) with a multidirectional orientation and great number of their intersections form. This leads to a considerable grain refinement and a continuous increase in the grains misorientation. In this case the dynamic polygonization and formation of subgrains precedes the dynamic recrystallization. A small size of these subgrains is mainly responsible for a high deformation hardening. The supposed mechanism of grain refinement after SPD at elevated temperatures consists in the gradual transformation of the originally formed subgrained dislocation substructure into a final structure with submicron and then nanocrystalline grain size. The lowering of the deformation temperature leads to a decrease in the intensity of dynamic softening processes (dynamic recovery) and, as a consequence formation of grains with smaller average size. The dynamic properties of various dislocation substructures, the acting slip systems and the deformation modes also change. All these aspects affect the properties of the finally formed microstructure.

The structure of alloy under all studied regimes of ECAP is shown in Figure 2.

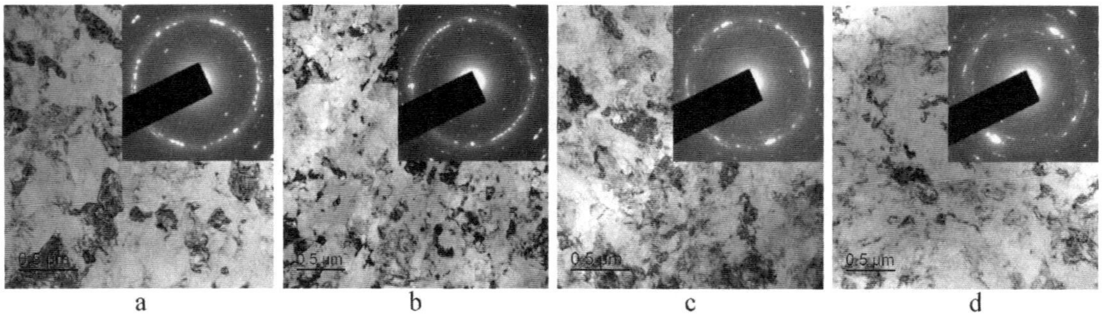

Figure 2. Microstructure of the TN-1 alloy subjected to deformation according to various regimes: (a) ECAP-20, (b) ECAP-3, (c) ECAP-7, and (d) ECAP-7 + 400 °C. Transmission electron microscopy: dark-field images, electron-diffraction patterns in the insets.

According to the results of the microstructural analysis, in the case of ECAP-20 at 450 °C, a mixed ultrafine grained/subgrained structure with high density of free dislocations is defined. The average linear size of the structural elements is 171±10 nm, therefore, this structure can be referred to a submicrocrystalline (Figure 2 a). The decrease in the deformation temperature to 400 °C and the exclusion of interpass heating lead to a refinement of structural elements. After ECAP-3, a mixed grained/subgrained structure with a large number of grains/subgrains with a size of less than 100 nm is produced. The average linear size of structural elements is 115±5 nm, i.e., it approaches the nanometer range from above. With an increase in the number of quasi-continuous passes to seven, the average size of structure elements does not significantly change, it rather stays at the boundary between the nanometer and submicrometer ranges (103±5 nm). The high density of free dislocations within the structural elements remains. Annealing after ECAP-7 at the deformation temperature (400 °C, 1 h) visually leads to a decrease in the dislocation density inside the grains/subgrains. Thus, after ECAP in 7 quasi-continuous passes, a mixed nanostructure of B2-austenite is produced. The size of structural elements of this nanostructure significantly exceeds the size, which was obtained by PDA after multi-pass cold rolling (40-80 nm), but the difference between completely recoverable strain values, obtained by PDA after cold rolling on a thin sample and ECAP on a bulk sample, is not so significant.

3.2. Comparison of mechanical and functional properties

In order to understand the main features of the structure, obtained after multi-pass cold rolling, and allow achieving the best combination of functional properties, such as recovery stress and completely recoverable strain, it is necessary to determine the conditions providing reduction of hindrances for reverse movement of shape recovery carriers. The first condition consists in the relatively high value of the austenite dislocation yield stress σ_y that serves as an approximate measure of the recovery stress value σ_r. The second condition is a relatively low value of the critical stress σ_{cr} for reorientation or formation of martensite. The maximum difference between these two values $\Delta\sigma = \sigma_y - \sigma_{cr}$ was obtained directly after multi-pass cold rolling and allow achieving an ultimately high combination of functional properties (completely recoverable strain of 8.0 %, and recovery stress of 1400 MPa) (Figure 3) [15].

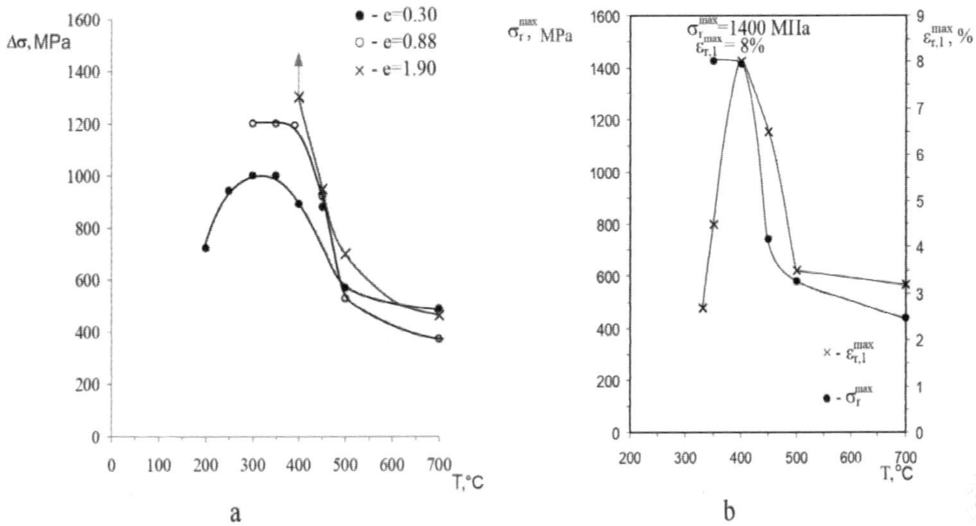

Figure 3. The values of $\mathbf{\Delta\sigma}$, $\boldsymbol{\sigma_r^{max}}$, $\boldsymbol{\varepsilon_{r,1}^{max}}$ after multi-pass cold rolling as function of the temperature of PDA. Arrow means that dislocation yield stress wasn't achieved during the tests [adapted from 10].

Formation of nanoscale grains with optimal degree of dispersion (40-80 nm) as a result of PDA at 400 °C after multi-pass cold rolling with accumulative true strain e = 1.9 increases the value of σ_y and the value of $\Delta\sigma$, but dislocation density in the obtained structure is low in comparison with the polygonized structure, when the high value of σ_y defined by high dislocation density. PDA at higher temperatures leads to the growth of austenite grains, consequent softening and degradation of functional properties. The structure with smaller grain size (less than 40 nm) after PDA at 350 °C, inhibits development of martensitic transformation and thus decreases the recoverable strain to 4.5 % and below. Nevertheless, the recovery stress remains at a very high level (1400 MPa) [15].

Equal channel angular pressing with the channel intersection angle of 120° at 400 °C for seven passes in the quasi-continuous regime allows achieving a mixed nanocrystalline and nanosubgrained structure with the average size of structure elements of 103±5 nm. As mentioned above, that is larger than the optimal range of 40-80 nm. Obtained structure is characterized by high dislocation density and the high ultimate tensile strength (1000-1150 MPa) in combination with the elongation to failure more than 20 %. The maximum completely recoverable strain in bending tests (9.5 %) was achieved after additional PDA at 400 °C for 1 h. Increase of the completely recoverable strain from 7.3 % after ECAP to 9.5 % after ECAP + PDA seems to be consequence of a decrease of the transformation yield stress and the retention of the high dislocation yield stress and, as a result, of an increase in their difference $\Delta\sigma$ [10].

3.3. Features of texture development during ECAP and rotary forging

The formation of crystallographic texture as a result of warm rotary forging (350-450 °C) and ECAP under different regimes described above has been studied. Partial DPFs {110} of rods deformed by warm rotary forging are shown in Figure 4. Partial DPF {110} of rod deformed by ECAP-20 with heating between passes up to 450 °C is presented in Figure 5a, and PPF {110} of rod deformed by ECAP-3 in quasi-continuous regime at the nominal temperature of 400 °C for three passes is shown in Figure 5b. All DPFs are presented for the section perpendicular to the axes of the rods.

IOP Conf. Series: Materials Science and Engineering **503** (2019) 012024 doi:10.1088/1757-899X/503/1/012024

According to the results obtained (Figure 4a), in the case of rotary forging at a temperature of 450 ° C, an axial tensile texture, at which the <110> axis is oriented along the direction of drawing of the rod, is formed in TN-1. Such a texture is typical for the deformation of the ordered crystals with a CsCl-type lattice along the slip systems {110} <001> [16]. When the deformation temperature is reduced down to 350-400 °C, an additional component with an axis <100> appears in the rod texture. Formation of this component can be connected with activation of other slip systems, for example, {001} <100>, as it was observed in the rolling of single crystals [17].

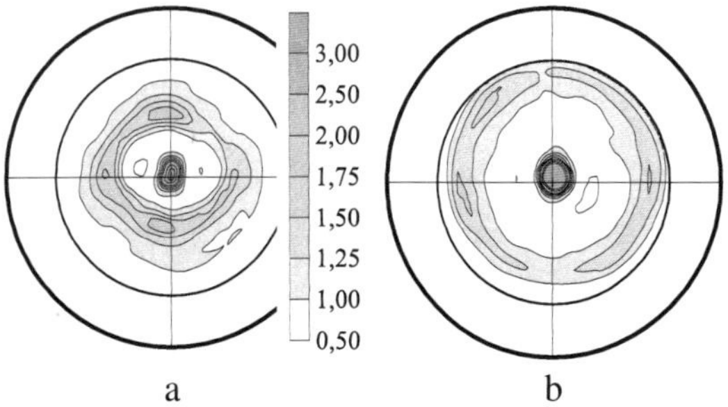

Figure 4. Partial DPFs {110} of rods, which were rotary forged at temperatures 350 (a) and 450 °C (b)

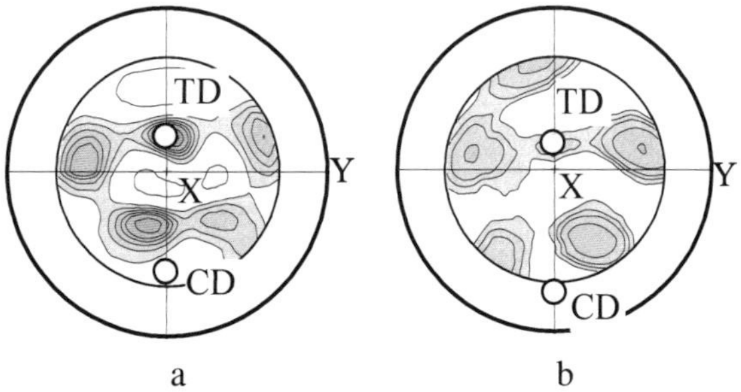

Figure 5. Partial DPFs of the TN-1 rods subjected to deformation according to various regimes: ECAP-20 at 450 °C (a) and ECAP-3 at 400 °C (b)

The partial DPFs {110} for different ECAP modes are shown in Figure 5. The outer axes of the rod after ECAP are denoted. The X axis is directed along the axis of the rod after deformation and displaced in the center of the stereographic projection. The Z axis is displaced at 120° relative to X-axis and characterizes the direction of movement of the rod at the beginning of the passage. The Y-axis is perpendicular to the plane of the intersecting channels directed along axes X and Z. The Z axis isn't shown in Figure 5. Earlier [17-18], it was shown that the ECAP texture can be represented as a rolling texture rotated about a transverse direction Y by some angle. The angle of rotation is determined by the angle of the channel intersection. The maximum shear plastic deformation develops in the rod when it passes the curved part of the channel. It is in this region that the rotation of the PPFs occurs, due to the reorientation of grains in accordance with the crystallographic mechanisms of plastic deformation acting there and the stressed state of the rod, given by the shape of the channels.

Let us consider deformation of TN-1 rod at the temperature 450 °C. As a result of 20 passes of ECAP, a clear crystallographic texture forms. It was shown in [18-19] that at each successive stage, a finite deformation texture that is characteristic of a stress state realized for this final stage can be formed, provided that the crystallographic mechanisms inherent in the material under consideration are not suppressed. This is due to the fact that the deformation at each stage exceeds 1. For forging at a temperature 450 °C, it is shown that crystallographic slip {110}<001> is activated, then in the case of rolling the ordered crystals with a CsCl-type lattice a {111}<110> texture is formed [16]. In accordance with the obtained results (Figure 5), it is this texture component that is rotated by an angle of about 100 ° relative to the first channel of the mold or Z-axis around Y-axis. The location of the TD and CD directions for the tool under consideration is shown in Figure 5.

In the case of ECAP at a temperature of 400 °C, the rolling texture {111}<110> still rotates and the increased scattering of texture is observed. However, the angle of rotation increases, which is due to a change in the coefficient of friction of the rod with the mold when the temperature changes. The observed scattering of the crystallographic texture can be associated with a change in the mechanisms of plastic deformation, which was also observed in the case of warm forging, as well as a result of grain refinement and activation of grain boundary slip. With the increase in the number of passes of ECAP, the texture maximum in PPF {110} approaches to the X axis.

Thus, the rods obtained by the warm rotary forging and ECAP in different modes are characterized by different crystallographic texture, the PDA also affects the final texture. Taking into account that the shape memory effect (SME) is orientationally dependent, the differences in the texture of the studied samples can affect the SME value.

4. Summary

1. The realization of the best combination of functional properties of Ti-Ni SMA may be explained by the peculiarities of mixed nanograined/nanosubgrained structure which forms as a result of annealing after severe plastic deformation. The increase in the yield stress due to the grain refinement down to a critical size of 40-80 nm in thin objects is comparable to the growth of the dislocation yield stress in bulk billets with a mixed nanograined/nanosubgrained structure with a high dislocation density after ECAP in the quasi-continuous mode and prevents plastic deformation by dislocation slip during deformation inducing SME. Besides, a positive effect of the post-deformation annealing is expressed in elimination of excessive work hardening and decrease of the transformation yield stress, which increases the completely recoverable strain.

2. It was found that when the temperature of the warm rotary forging of the TN-1 alloy is reduced from 450 to 350-400 °C, along with slip systems {110} <001>, operating at 450 °C, additional plastic deformation systems {100} <001> are activated.

3. In the case of ECAP-20 with intermediate heating up to 450 °C, the same slip systems act as in the case of rotary forging, which determine the formation of a characteristic rotated rolling texture at ECAP. The reduction in the pressing temperature and the absence of intermediate heat between the ECAP passes contribute to scattering of the crystallographic texture while maintaining the predominant texture component inherent in the deformation of ECAP-20.

Acknowledgements

The present work was carried out with financial support from the Ministry of Science and Higher Education of the Russian Federation (State Task No.11.1495.2017/4.6) and RFBR (research project № 18-08-01193 A).

References

[1] Brailovski V, Prokoshkin S, Terriault P and Trochu F 2003 *Shape Memory Alloys: Fundamentals, Modeling and Applications* (Montreal: ETS) p 844
[2] Jani J M, Leary M, Subic A, and Gibson M A 2014 *Materials & Design* **56**, 1078-113
[3] Resnina N and Rubanik V 2015 *Shape Memory Alloys: Properties, Technologies, Opportunities* (Pfaffikon: Trans Tech Publications) p 640
[4] Valiev R, Estrin Y, Horita Z, Langdon T, Zehetbauer M and Zhu Y 2006 *JOM* **58** 33-9
[5] Sabirov I, Enikeev N, Murashkin M and Valiev R 2015 *Bulk nanostructured materials with multifunctional properties* (New York: Springer) p 118
[6] Lotkov A, Grishkov V, Kashin O, Baturin A, Zhapova D, and Timkin V 2015 *Mater. Sci. Found.* **81-82** 245-59
[7] Khmelevskaya I, Komarov V, Kawalla R, Prokoshkin S, and Korpala G 2017 *J. Mater. Eng. Perform.* **26** 4011-19
[8] Khmelevskaya I, Komarov V, Kawalla R, Prokoshkin S and Korpala G 2017 *Mater. Today: Proc.* **4** 4830-35
[9] Stolyarov V V, Prokofiev E A, Valiev R Z, Prokoshkin S D, Dobatkin S V, Trubitsyna I B, Khmelevskaya I Yu and Pushin V G 2005 *Phys. Met. Metallogr.* **100** 2703-14
[10] Inaekyan K E, Prokoshkin S D, Brailovski V, Demers V., Dobatkin S V, Tatyanin E V and Bastarache E 2006 *Mat. Sci. Forum* **503-504** 597-602
[11] Prokoshkin S D, Khmelevskaya I Yu, Andreev V A, Karelin R D, Komarov V S and Kazakbiev A M 2018 *Mat. Sci. Forum* **918** 71-6

IOP Conf. Series: Materials Science and Engineering **503** (2019) 012024 doi:10.1088/1757-899X/503/1/012024

[12] Khmelevskaya I Yu, Prokoshkin S D, Trubitsyna I B, Belousov M N, Dobatkin S V, Tatyanin E V and Prokofiev E A 2008 *Mat. Sci. Eng.* **481-482** 119-22

[13] Khmelevskaya I Yu, Karelin R D, Prokoshkin S D, Andreev V A, Yusupov V S, Perkas M M, Prosvirnin V V, Shelest A E and Komarov V S 2017 *Phys. Met. Metallogr.* **118** 279–87

[14] Perlovich Y, Isaenkova M and Fesenko V 2016 *IOP Conf. Series: Mat. Sci. Eng.* **130** 012055

[15] Brailovski V, Prokoshkin S D, Khmelevskaya I Yu, Inaekyan K E, Demers V, Dobatkin S V and Tatyanin E V 2006 *Mat. Trans.* **47** 795-04

[16] Vishnyakov Ya D, Babareko A A, Vladimirov S A and Egiz I V 1979 *Theory of texture formation in metals and alloys* (Moscow: Nauka) p 329 (in Russian)

[17] Perlovich Yu, Isaenkova M, Fesenko V, Dement'eva T and Gol'tsev V 2013 *Rus. Met. (Metally)* **4** 300-3

[18] Perlovich Yu, Isaenkova M, Fesenko V, Grekhov M, Alexandrov I and Beyerlein I J 2006 *Mat. Sci. Forum* **503-504** 853-8

[19] Perlovich Yu, Isaenkova M, Fesenko V, Grekhov M, Seng-Ho Yu and Sun-Keun 2006 *Mat. Sci. Forum* **503-504** 859-64

Analysis of the conditions of deformation's formation in lead selenide nanolayers

A M Pashaev[1], O I Davarashvili[2], M I Enukashvili[2], Z G Akhvlediani[2,3], L P Bychkova[2], R G Gulyaev[2], M A Dzagania[2,4] and V P Zlomanov[5]

[1] National Aviation Academy of Azerbaijan
[2] Iv. Javakhishvili Tbilisi State University, Georgia
[3] E. Andronikashvili Institute of Physics, Georgia
[4] P. Melikishvili Institute of Physical and Organic Chemistry, Georgia
[5] M. Lomonosov Moscow State University, Russia

E-mail: omardavar@yahoo.com

Abstract. The results of reservation of strained state in PbSe layers grown on KCl, BaF_2, NaCl, CaF_2 substrates during several years and revealing of the peculiarities of two-stage growth of layers, the creation of supercritical structures, the study of the broadening of elastic deformation at layers doping with impurities of varying valence show that it is reasonable to regulate in detail the strains (deformations) in nanolayers. For definite groups of layers with the thicknesses in the range of <100 nm and 100-200 nm, as well as at not high – 1 nm/s and high ≥10 nm/s rates of growth there appear the new means to control the deformations in PbSe nanolayers.

1.Introduction

For application of the supercritical strained PbSe layers at "negative" pressures suggested earlier in [1] for the solution of new technical problems, i.e. the development of high-temperature and high-sensitive photodetectors [2], it is interesting to consider the problems of interconnection of the thickness of layers and the rate of their growth with the deformations taking place in them [3]. PbSe layers were fabricated on BaF_2, NaCl, CaF_2 and KCl substrates with different mismatch. The special interest was caused by PbSe/KCl structures, as in this case maximal deformation of stretching was realized that will find new application in IR optoelectronic devices. The choice of the method of molecular epitaxy with a "hot-wall" [4-7] gives a possibility to control the rate of layer growth and to show the importance of this parameter for obtaining the maximum deformation also at the maximum thickness of the layer. The development of the method of controlling the growth of layers will additionally allow to define the peculiarities of their growth [8-9] and to investigate in detail the properties of layers at each step of the two stages of the growth. The investigation of these problems is the subject of the given work.

2. Experimental methods

For the planned investigations the layers of PbSe were obtained by the method of molecular epitaxy with a "hot-wall". The flux of molecules to the KCl (100) substrate is formed by heating of polycrystalline PbSe in the range of temperature 450-510^{0}C. KCl substrate was heated in the range of temperature 240-320^{0}C. The rate of growth was controlled both by the change of the temperature of the source of epitaxy and by the distance from the opened edge of quartz ampule with the source to the substrate in the range of 1-20 mm. The lattice constant of layers and their thickness, and the half-width of X-ray line were determined by registration of X-ray diffraction patterns at the reflection from (200),

(400) and (600) planes [10]. Deformation ε in PbSe layers was determined via the lattice constant of strained and unstrained layers of PbSe α_l and α_{PbSe} as $\varepsilon=(\alpha_l-\alpha_{PbSe})/\alpha_{PbSe}$. Electrical properties (concentration of current carriers and their mobility) of the layers were determined by Van der Paw method.

3. Results and discussion

PbSe layers with <200 nm thickness were mainly investigated.

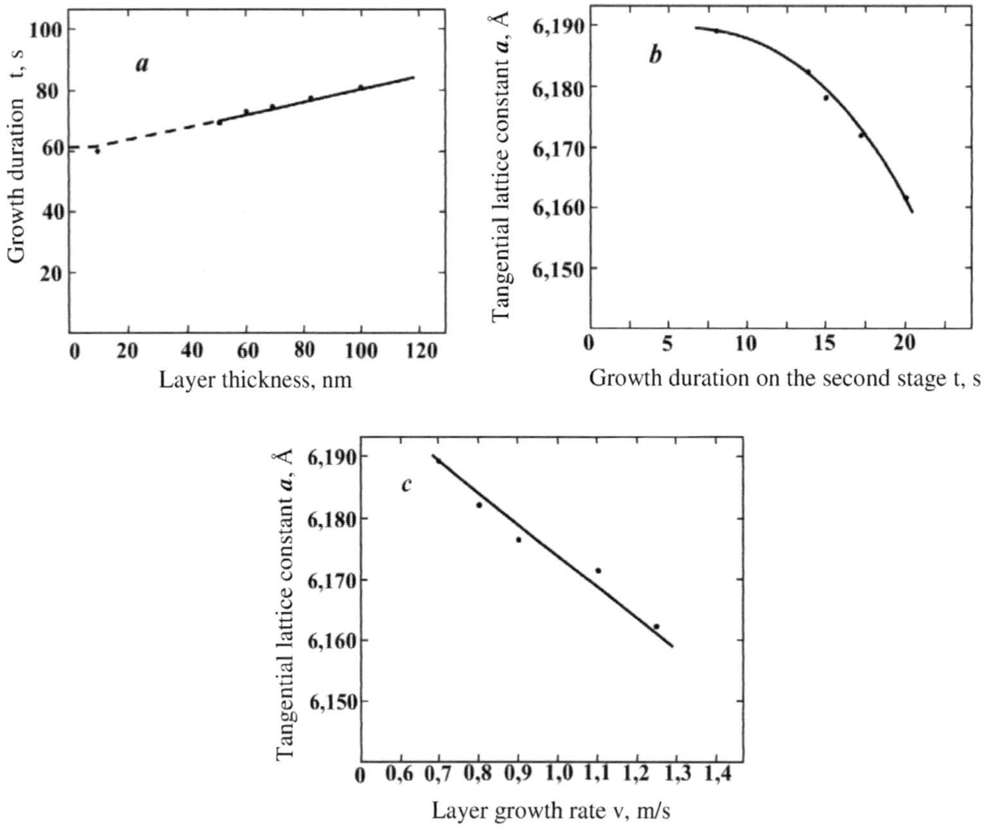

Figure 1. Dependence of tangential lattice constant and layer thickness on growth rate and duration (*a*,b,c)

Figure 1α presents the dependence of the thickness of layers (with <100nm thickness) on the time of growth for specific regimes: at the temperatures of source 470^{0}C and of KCl substrate - 300^{0}C (320^{0}C) and at the distance from the open edge of quartz ampule to the substrate 12 mm. On the basis of such five layers in the row SL 742 – SL 763 the data are presented in the table. The duration of growth (t), depending on the thickness of layers, is limited to 60 sec at the decrease of the thickness of layer to 0, more exactly, to 10 nm, being the sensitivity of the X-ray method of measuring a thickness. Probably, the time 60 s for the specific conditions is the time of "delay" – corresponds to the first stage of growth. On the first stage, emerging and merging of nuclei islands, on the second stage – layerwise growth, their duration depends on the conditions of growth (temperatures of source and substrate, distance between ampule and substrate, mismatch). If we suppose that within the thicknesses of 100 nm, the second stage

IOP Conf. Series: Materials Science and Engineering **503** (2019) 012025 doi:10.1088/1757-899X/503/1/012025

of layerwise growth starts not at the thickness above 10 nm, but even at higher thickness, i.e. 40-50 nm, as the estimations show, this will cause the change of the full rate of growth – 0.1 nm/s. The similar investigations at the distance between the open edge of quartz ampule with the epitaxy source and the substrate equal to 1mm, 2.5 mm and 20 mm made it possible to determine the duration of the first stage of emerging and merging of nuclei islands – 2, 9, 100 s respectively [11]. For the layers SL 742-763 distance between ampule and substrate was 12 mm, for the layers SL 781-284 – 2.5 mm, for the layers SL 613, 616 – 1mm.

Table. Data on the conditions of growth and characteristics of strained PbSe layers on the KCl(100) substrate

N	Layer	Temperature of the source T,⁰C	Temperature of the substrate T,⁰C	Growth duration t, s	Layer thickness d, nm	Tangential lattice constant *a*, Å	Layer growth rate v, nm/s	X-ray dif. line half-width Δ, min	Relative mismatch -deformation $\varepsilon = \frac{a_l - a_{PbSe}}{a_{PbSe}} \cdot 10^2$
1	SL-742	470	300	80	101	6,162(3)	1,25	17	0,59
2	SL-758	470	300	77	83	6,172(1)	1,1	17	0,75
3	SL-756	470	300	75	68	6,178(4)	0,9	18	0,83
4	SL-751	470	320	74	59	6,185(8)	0,8	19	0,95
5	SL-763	470	300	67	48	6,188(2)	0,7	20	1,01
6	SL-781	470	240	10	35	6,210(2)	3,5	22	1,38
7	SL-783	470	240	13	98	6,145(6)	6,7	24	0,29
8	SL-215	450	240	300	189	6,129(7)	0,6	21	0,05
9	SL-578	470	240	30	181	6,149(0)	6,1	25	0,38
10	SL-284	470	240	15	175	6,136(7)	11,7	31	0,16
11	SL-613	510	240	11	204	6,138(8)	18,7	33	0,21
12	SL-616	510	240	19	435	6,131(9)	22,8	34	0,10

Such analysis allowed to determine the rate of layer growth at the second stage, with the exception of the thickness equal to 10 nm and the time of delay at the first stage equal to 60 s, as well as the rate of growth at the second stage ranged within 3.5-5.4 nm/s and the average rate was 4.5±1 nm/s. Thus, one can think that the deformation in layers connected with tangential lattice constant will be changed according to the growth duration at the second stage (the average rates of growth at the second stage are taken to be the same for all layers).

Figure 1b shows the dependence of tangential lattice constant on the growth duration at the second stage. It is seen that at the increase of growth duration from 7 to 20 s the tangential lattice constant decreases from 6.190Å to 6.160 Å. Within such an approach one can determine the dependence of tangential lattice constant on the full rate of layer growth. This rate of growth can be determined in two ways: by the relation of layer thickness to the growth time, and by the share of growth rates contributed at the first and the second stages.

For instance, the full average rate of growth can be estimated as the relation of growth rate 4.5 nm/s at the second stage (for layer SL – 742) to the value 0.9/0.25 – the resultant rate makes 1.25 nm/s and corresponds to the rate calculated from the relation 101(d)/80(t) = 1.25nm/s. It is important to note that 60s as a delay time is rough value (real is slightly more, Figure 1α). This is a result that in estimation was considered average value of the growth rate. The dependence of tangential lattice constant on the growth rate, presented in Figure 1c shows the decrease of the full growth rate with the increase of tangential lattice constant. This is connected with the fact that such result is achieved in the layer with the thickness of 48 nm (SL-763) and at the second stage a significantly less time is spent for the growth of this layer than in the case of SL-742 layer, when the full average rate of growth increases because of more contribution of much higher rate at the second stage (with the increase of the rate 90 nm layer increases).

The following two layers on the table: SL-781 and SL783 (with the thicknesses ≤ 100 nm) are obtained at much higher growth rate at the expense of the decrease of the distance from the open edge of quartz ampule with the source of epitaxy to the substrate down to 2.5 mm. By way of example of these two layers it is seen that with the increase of growth rate, the layer SL-781 with the thickness of 35 nm is growing without transferring the elastic energy to the dislocations and its lattice constant reaches 6.210(2) Å and the deformation – $1.38x10^{-2}$. For the layer SL-783 the thickness reaches 98 nm and, in this case, on the contrary, the nonstoichiometric defects have no time to retard the motion of dislocations and, correspondingly, the loss of layer energy increases and the lattice constant decreases to 6.145(6) Å, and the deformation decreases to 0.89×10^{-2}.

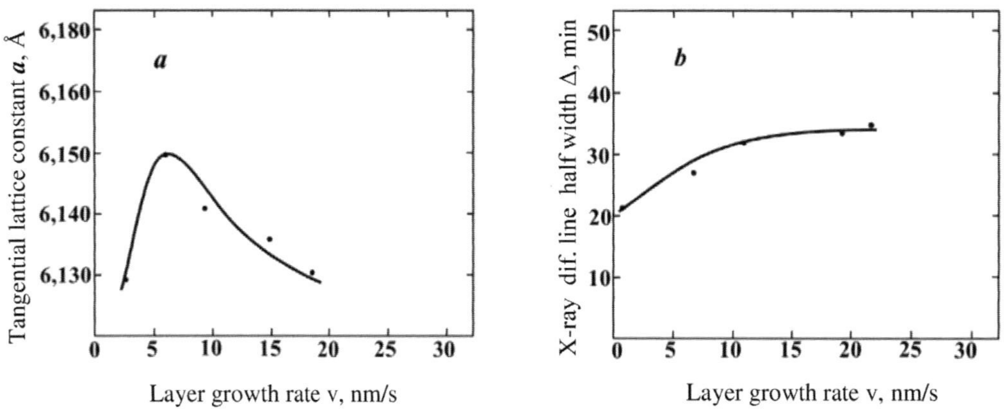

Figure 2. Dependence of tangential lattice constant (*a*) and X-ray dif. line half-width(b) on growth rate

For the second group of layers SL-215, 578, 284, 613, 616 (mainly, with the thicknesses 100-200nm) the dependence of tangential lattice constant on the growth rate has of nonmonotonic character: at the full rate of growth 6nm/s it passes through the maximum with the slow decrease at the increase of growth (Figure 2*a*). With the increase of thickness of layers in the strained state, their elastic energy increases so that at the concentration of nonstoichio-metric defects 10^{18}- 5×10^{18} cm^{-3} is not sufficient to create barriers against the shifting of formation dislocations. This is well shown by SL-215 layer, when the lattice constant makes 6.129 (2) Å at the little rate of growth. With the increase of growth rate there appears the tendency of disordered distribution of dislocations, and the cloud of nonstoichiometric defects around the dislocations can retard their motion and the lattice constant reaches 6.149(0) Å at the 6 nm/s growth rate. With the further increase of growth rate, the breaks between the clouds appear, and on the other hand, with thickening of layers the dimensions of subgrains increase in them, and with the increase of growth rate - the mismatch on their boundaries increases as well. Conditions of the growth for all layers were found for receiving single-crystal layers.

The critical dimensions connected with such mismatch become less than the dimensions of subgrains themselves and besides the boundary substrate – layer there appears an additional channel for relaxation of strains. For the layer with 204 nm thickness (SL-613) the lattice constant decreases to 6.138(8) Å at the rate of growth 18.5 nm/s. Figure 2a shows how the mismatch between subgrains increases at the increase of growth rate for the mentioned layer, i.e. the half-width of X-ray diffraction line reaches 33'(Figure 2b) .

In the case of layer SL-616, the mutual influence of mismatch between the subgrains and of breaks in clouds of nonstoichiometric defects on the decrease of lattice constant down to 6.131(9) Å at the growth rate 22.8 nm/s is illustrated. The deformation, probably, is kept up to the thickness 500 nm.

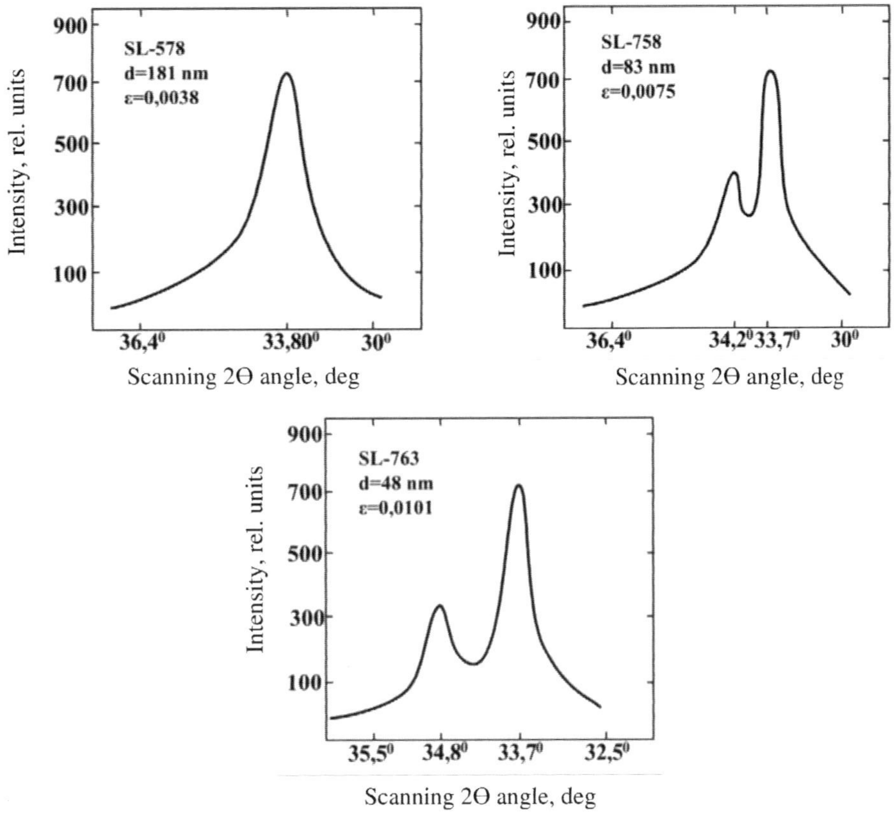

Figure 3. Diffractograms Θ-2Θ scanning of layers with different thicknesses

It is interesting to note, that comparing the data on layers SL-284 and SL-613, some increase of the lattice constant in the second case is connected with the detection of pores on the edge of SL-284 layer and its real thickness is higher than the mentioned above determined by X-ray method. The pores were revealed by optical microscopy.

Thus, for the second group of layers with high thicknesses the tangential lattice constant decreases and the increase of growth rate leads to its additional decrease because of less influence of nonstoichiometric defects (breaks in clouds) and because of relaxation of strains in subgrains. In general, at "negative" pressure, the deformation developed in the layer with 35 nm thickness, makes $1.38 \cdot 10^{-2}$, and for the layer with thickness, i.e. 70 nm – $0.8 \cdot 10^{-2}$.

Figure 3 presents the X-ray diffractograms 2θ of scanning from plane (200) for three PbSe layers of different thicknesses. For the thickness <100 nm the spectra are splitting and an additional peak appears at larger 2θ angles. The splitting between the main line and that of appeared again increases with the decrease of layer thickness – the increase of deformation. Such difference can be used for estimation of deformation in layers. As the estimations show, at such deformations the pressure of 10-15 kbar is developed and the texture of tetragonal phase appears in the layer.

The investigations of the spectra of optical absorption of the strained PbSe layers plotted according the transmission spectra showed that the tetragonal phase in deformed PbSe layer leads to the appearance of a new level in the conduction band and to the additional absorption connected with it alongside with the main band-to-band absorption [12]. This peculiarity was also caused by strained nanolayers of lead selenide.

4. Conclusions

The influence of layer thickness and of the growth rate on the tangential lattice constant is considered for the thicknesses <100nm and 100-200 nm at the growth rate ~1 nm/s and ≥10nm/s. The peculiarity of the growth for the layers of <100nm and at the growth rates ~1 nm/s is that the tangential lattice constant can be controlled depending on the duration of growth at the second stage – the rate of shifting of nonstoichiometric defects is slightly slower than the rate of growth. With the increase of growth rate for the same thicknesses of layers, the rate of shifting of nonstoichiometric defects is already noticeably less than the rate of growth and the rate of shifting of dislocations and the lattice constants begin to decrease.

For the thicknesses of layers within 100 -200 nm, the dependence of tangential lattice constant on growth rate is of nonmonotonic character. In this case, with the increase of growth rate the disordered distribution of dislocations takes place and now they are retarded by the cloud of nonstoichiometric defects. The dimensions of subgrains and mismatch between them are increased leading to the additional relaxation of strains. In view of the fact that the rate of the transfer of elastic energy increases with the increase of layer thickness, it is reasonable to decrease the rate of delivery of evaporating molecules to the substrate, e.g., by a smooth increase of the distance from the open edge of quartz ampule with the source of epitaxy to the substrate. Wide possibilities of the regulation of the conditions of growth were allowed to choose two parameters: thickness and growth rate of the layers for the analysis of the process for reaching maximal deformations.

In nanolayers at the increasing deformations there appears a tetragonal phase leading to the appearance of a new level in the conduction band and to the additional absorption in the optical spectrum.

References

[1] Pashaev A M, Davarashvili O I, Enukashvili M I, Akhvlediani Z G, Bychkova L P, Gulyaev R G, Dzagania M A and Zlomanov V P 2017 Supercritical lead selenide layers at "negative" pressures *Bull. Acad. Sci. Georgia* **43** 2 179

[2] Pashaev A M, Davarashvili O I, Enukashvili M I, Akhvlediani Z G, Bychkova L P, Gulyaev R G, Dzagania M A and Zlomanov V P 2017 Investigation of the strained nanolayers of IV-VI semiconductors on dielectric substrates *Bull. Acad. Sci. Georgia* **43** 1 24

[3] Pashaev A M, Davarashvili O I, Enukashvili M I, Akhvlediani Z G, Gulyaev R G and Dzagania M A 2015 Variations of the parameters of PbSe nanolayers with change of their technology *IOP Conf.Ser Materials Science and Engineering* **77** 012017

[4] Doh K and Preier H 1975. Properties of $PbS_{1-x}Se_x$ epilayers deposited on PbS substrates by "hot-wall" epitaxy *Journal of Material Science* **10** 1360

[5] Lopez-Otero A hot –wall epitaxy 1978 *Thin Solid Films* **49** 1

[6] Viatkin K V, Shotov A P and Ursaki V V 1981 Thin layers $PbS_{1-x}Se_x$ grown by "hot-wall" epitaxy *Inorganic Materials* **17** 1 24

[7] Pashaev A M, Davarashvili O I, Enukashvili M I, Akhvlediani Z G, Bychkova L P, Gulyaev R G 2013 Epitaxial lead selenide layers over a wide range of their thickness on dielectric substrates *J. Mater. Sci&Eng* B**3**(2) 97

[8] Palatnik L S, Sorokin R K, Zozulya L P 1969 The mechanism of growth of isolated particles at epitaxy of lead chalcogenides on NaCl and KCl. *Solid State Phys* **11** 5 1265

[9] Pashaev A M, Davarashvili O I, Enukashvili M I, Akhvlediani Z G, Gulyaev R G, Zlomanov V P 2013 Nanolayers of lead selenide on potassium chloride substrates: specific features of formation *J. Mater. Sci&Eng* **A3**(4) 235

[10] Pashaev A M, Davarashvili O I, Enukashvili M I, Akhvlediani Z G, Bychkova L P, Gulyaev R G, Dzagania M A 2017 Lattice constant as indicator of the technology and properties of IV-VI semiconductors *Bull. Acad. Sci. Georgia* **40** 2-3 11

[11] Pashaev A M, Davarashvili O I, Enukashvili M I, Akhvlediani Z G, Bychkova L P, Gulyaev R G, Dzagania M A, Zlomanov V P 2016 Investigation of the strained lead selenide nanolayers *Global Journal of Engineering Science and Research Management* **3**(6) 56

[12] Pashaev A M, Davarashvili O I, Enukashvili M I, Akhvlediani Z G, Bychkova L P, Gulyaev R G, Dzagania M A, Zlomanov V P 2016 Study of the optical characteristics of epitaxial $PbS_{1-y}Se_y$ layers *Global Journal of Engineering Science and Research Management* 3(2) 46

IOP Conf. Series: Materials Science and Engineering **503** (2019) 012026 doi:10.1088/1757-899X/503/1/012026

Optimization of the enzyme power source for a nano drug delivery system fuelled by glucose in blood plasma

S Naidoo[1], L Thage[1], Q Ying[2], S Vallie[1] and G Vaivars[3]

[1]University of Stellenbosch, South Africa,
[2] University of Cape Town, South Africa,
[3]Institute of Solid State Physics, Department of Chemistry, University of Latvia, Latvia

Email: gasennaidoo@sun.ac.za

Abstract. A unique in vivo electrical pulse generator to improve membrane permeability for drugs and simultaneously facilitate self-powered nano devices for nano drug delivery systems (NDDS) was identified. The use of an unsupported biological catalyst component of the power supply was aimed at the NDDS instead of a conventional membrane electrode assembly (MEA). Self-powered carriers of drugs and prodrugs with improved controlled release capability to target areas using substrate available in biological matrices such as glucose in blood is envisaged. The experimental application implemented prototype designed chambers allowing the entry of premixed precursors and low ohm resistance due the absence of diffusion layers and optimised open circuit voltage (OCV). This would also minimise poisoning and rupturing of the proton exchange membrane (PEM). The model uses the isothermal experimental design (37°C) parameter and the glucose is partly oxidised prior to entry and mostly oxidised at the surface of the proton exchange membrane (PEM). The experimental model used a residence time instead of the usual flow rate. The power was notably high for short periods due to the absence of carbon supported diffusion layers. The findings included low levels of glucose and glucose oxidase (GOx) are needed for OCV optimisation.

1. Introduction

Nature has over time established a chemical energy supply from biomass sources using complex metabolic pathways. Scientists are using this information to optimise in vitro power cells to power sustainable in vivo devices and other stationery device applications [1]. Where blood plasma contains approximately 90% water, drug dosage (concentrations) undergo a dilution effect coupled with an efficient blood circulatory system (systemic) without the efficacy of targeting may lead to poor therapeutic regimens and possible mutations where bacterial infections like Tuberculosis (TB and MTB) occur leading to extreme drug resistance (XDR) and deteriorating patient health where inadequate dosing is encountered. Polar drugs are more likely to be excreted via the urinary pathway further lowering the bioavailability of the intended therapeutic dose reducing efficacy. Multiple drug regimens and crushed drug (tablets) administration, to treat co–infections, are possibly affected by drug-drug interaction due to differences in the pKa (weak acid / base in pH buffered matrices)) chemistry could possibly compromise efficacy. A biologically compatible NDDS in physiological matrices is essential in maintaining the optimised pharmacokinetic (PK) levels ensuring the therapeutic drug dosage removes all symptoms and infection. The complex biological homeostasis energy balanced systems can counter the fluctuations in glucose as the precursor is consumed and the water by-product is formed. In a biological matrix the fuel source is placed in the anode chamber of the cell in the presence of a potentially biologically active catalyst (enzyme). At the anode, the enzyme (protein) oxidises the fuel source (glucose) to produce carbon dioxide and water. This chemical process does not only yield carbon dioxide and water but also yields electrons and protons.

$C_6H_{12}O_6 + H_2O \rightarrow C_6H_{12}O_7 + 2e^- + 2H^+$ .... (1)

The following reactions are seen to happen in a glucose fuel cell [2, 3]:

At the cathode, the electrons also flow to the positive terminal also known as the cathode through a conductive bridge for the completion of the cell. The current in this case flows from the negative terminal to the positive terminal [6].

$\frac{1}{2}O_2 + 2e^- + 2H^+ \rightarrow H_2O$... (2)

Overall cell reaction:

$C_6H_{12}O_6 + \frac{1}{2}O_2 \rightarrow C_6H_{12}O_7$... (3)

Here the biological power source is defined as a conversion mechanism of chemical energy stored in bio-molecule fuel into electrical energy released by the catalytic activity of enzymes. A substrate in this process is consumed and the electrons are produced from anode glucose catalytic GOx reaction. The electrons move to the negative terminal of the cell also known as the anode to the cathode electrode depending on the open cell potential efficiency. The optimised theoretical power cell potential for glucose as a fuel source is approximately 1.256 V [2, 3]. Owing to over-potential losses this will typically show operating glucose power source cells operate at an open circuit potential (OCV) of 0.5–0.8 V.

The model assumes that the glucose and hydrogen ion transport through the cell occurs through a membrane that is temperature dependant. The glucose fuel cell domain was divided into 3 compartments namely - the anode premix chamber (APC), the proton exchange membrane and the cathode premix chamber (CPC). The schematic of the fuel cell is shown in Figure 1 with the dimensions. It is assumed that the membrane is impermeable to glucose but permeable to hydrogen ions. Thus, the concentration of the glucose is modelled only until the membrane. The reaction of the glucose to form hydrogen ions is assumed to occur in the APC, in presence of the enzyme, GOx. The unreacted glucose and the hydrogen ions diffuse across the cell to the cathode. The unreacted glucose is assumed not to react at the cathode. The model assumes that the transport of glucose is solely by diffusion and convection. The effect of the potential drop across the glucose fuel cell on glucose molecules is considered to be negligible (glucose is not a charged species). The potential drop across the membrane is considered to be a significant factor in the transport of hydrogen ions across the membrane. The effect of potential drop on the transport of hydrogen ions across the enzyme layer and the cathode catalyst layer is considered negligible given the thickness 183µm.The electrode biofouling can also contribute to the potential drop compounded by non-porous electrodes and unbound enzymes [4].

2. Experimental Methods

A scale-up power cell prototype model, Figure 1, was used as the power cell included an anode chamber contained GOx and glucose. The anode chamber contained neither facilitator nor mediator at temperature ranging from 36.9 to 37.1°C. The anode chamber was degassed using a nitrogen cylinder (N_2) fitted with a regulator, the inert gas was flowed through the system for 2 hours. The buffer used was a phosphate buffer to facilitate the ion migration and pH stability. The assumed Gibbs energy was -97.5kJ.mol^{-1} [8]. The electrodes used were carbon graphite rods, 1 per chamber (4 mm carbon rod electrode diameter; 50 mm length with only flat 4mm end immersed) where the electrode reactions are controlled by the enzyme kinetics [7].

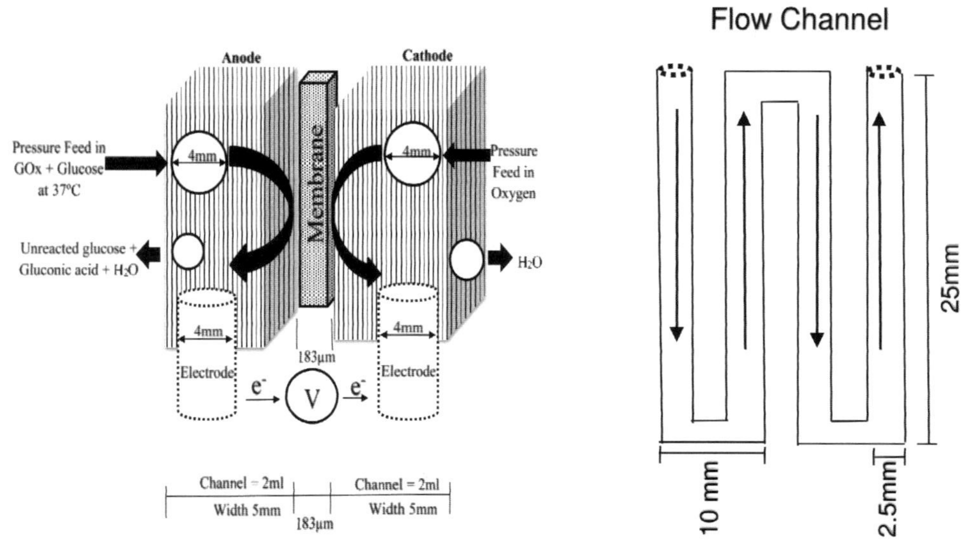

Figure 1: Power cell (EFC) with 2 electrodes and glucose / GOx feed in absence of diffusion layer

Electrical circuit was connected using insulated conductive and chrome plated electrically conductive crocodile clip connections. The clip connections were not in contact with the solutions in either chamber. The experiments were conducted under biological temperatures conditions and ranged from 229 to 331 K. The electrode area was calculated as follows.

$$A = 2\pi r^2 + 4(L_1 \times B_1) + 2(L_2 \times B_2) \quad \text{.......} \quad (4)$$

The sulfonated tetrafluoroethylene based fluoropolymer-copolymer Nafion® [6], was incorporated as it displayed sufficient stability previously [6]. The pH influence on open circuit voltage (vs. E) was measured by multi-meter (software) and Metrohm pH meter in 0.001M HCl (Kimix batch no. 264). The electrochemical activity (ECA) was studied using software attached to a sensitive multimeter. The basic mechanism of the electricity production in this cell was to allow the freely available glucose to react with glucose oxidase and co-factors in an aerobic environment. The glucose reacted with water to produce carbon dioxide, protons and electrons, Equations 1-3.

3. Results and Discussion

The influence of enzyme and substrate concentration was investigated as a redox relationship to produce hydrogen ions contributing to the OCV. The physical measurable components included the hydrogen ion concentration influencing the current density, electron and ionic charge, electrical conductivity and electrical field. The curent density is one dimensional and plotted on the x-axis, Figure 2a.

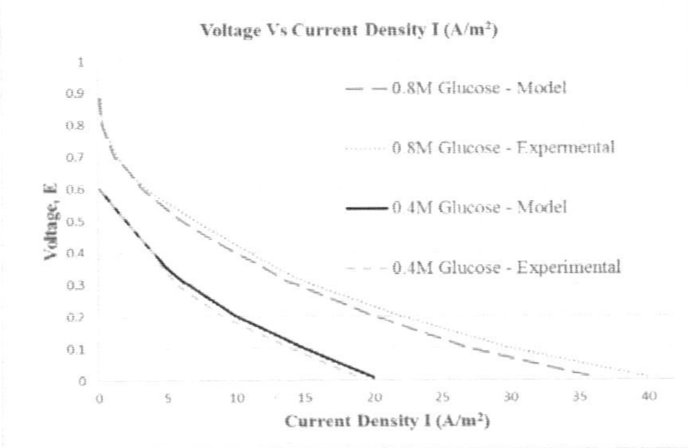

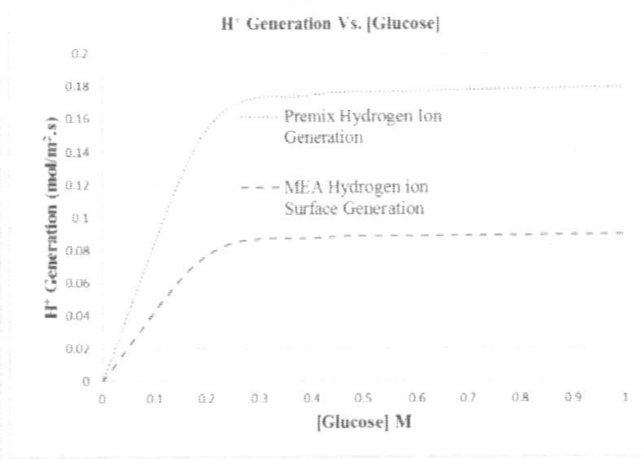

(a) **(b)**

Figures 2 (a) and (b): Modeled Polarization and comparative experimental data for varied glucose concentrations and hydrogen ion generation.

The divergence of current density depended on the hydrogen ion formation being relative to the oxidation and reduction (consumption of H^+) processes. The hydrogen ion generation occurred mainly at the anode compartment and the consumption, assuming membrane proton conductivity was optimal, was noted to occur in the cathode compartment. The hydrogen ion diffusion coefficient in water was accepted as $4.5x10^{-9}$ m^2/s.

Proton flux towards the cathode have significantly influenced the OCV. Phosphate buffer optimisation of pH dependant flux was seen to improve the potential difference at biological pH7. The flux of protons through the membrane between anode and cathode solutions is expressed as a function of the mass transfer coefficient and assuming a constant pH in the cathodic solution. The kinetic enzyme reaction rate coefficient was estimated as $1.6x10^3$ s^{-1} and the hydrogen ion surface generation (H^+_s) was calculated at $0.088mol/(s.m^2)$ for the MEA.

$$H^+_s = \frac{[E][G]\,2k_{cat}}{[G]+k_m} \quad \ldots\ldots (5)$$

Equation (5) where the concentrations for enzyme and glucose are estimated as $2.8x10^{-5}$ and 1.0M respectively for OCV approximately 0.45V when a membrane electrode assembly (MEA) is used, however a premix, Figure 2(b) can produce a shorter yet >2 fold power increase. There was a noticeable power surge and power spikes exceeding 42 A/m^2 recorded. The power discontinuity could be due to slow diffusion rates across the water and the PEM medium inhibiting the charged particle gradient formation affecting the OCV and depletion parameter.

4. Conclusion

The drug permeability enhancement from a NDDS with polar and non-polar drug carrying capability, in the lipophilic form and with a carrier that can enhance permeability by means of *in vivo* generated electroporisity pulses as a possible mechanism to supply pulses to the gated channels and possibly create new temporary transport channels has been supported with this research. With the ability to affect the capacitance capability of the bi-lipid membrane and the ion movement, the power source can stimulate the absorption of molecules to either support the sustained potential difference across the membranes or stimulate transport of and when necessary block potassium (K^+), calcium (Ca^{2+}) or sodium ions (Na^+) during the cells natural electrochemical operations.

Figure 2a, model fit for 2 glucose concentrations (0.4M and 0.8M) are in support of varying in vivo glucose levels to change the ECA power output. The varied concentrations support the introduction of a control mechanism that can regulate power and subsequent pulse frequency and intensity. Insulin level regulation will be monitored in future as a regulatory mechanism for ECA power outputs (influencing permeability) to adjust glucose and subsequent OCV levels.

Figure 2b, the premix power was higher than that of the MEA however short lived. Increases in potential were noted with increases in the glucose and enzyme (GOx) concentrations. OCV logarithmic dependence was evident in increased GOx concentrations that is relative to buffer concentrations. The voltage increased where the buffer concentration decreased.

NDDS design qualification criteria will influence the drug-transport efficiency in tissue once released from the carrier, where neutral drug forms are considered favoured in the pharmacological context being lipophilic. The buffer role and significance was highlighted during the proton formation and transport, as the proton formation can lower the pH however increased buffer concentrations have been noted to lower the OCV. The maintenance of the pH will become highly significant depending on the PKa of the drug in the carrier. The carrier (possibly a Micelle) will ideally be polar with the capability to carry non-polar drugs. Drugs which are considered weak acids will easily maintain their neutral configuration in a matrix pH below their PKa values. There may be a need for a compartmentalised power source and storage cell during the courier of weak bases as the pH needed is above the PKa to maintain the neutral electronic status of the molecules. The substrate oxidation and proton generation may form ionisation of the drugs considered weak base. The neutral glucose molecule can be oxidised however a passing glucose molecule in plasma can be secured within the NDDS carrier by phosphorylation ionisation, with a charge of -1.8 easily binding to the enzyme by electrostatic interaction.

The optimized model will include glucose receptors and binding sites ensuring the constant supply of glucose in a porous electrode anode chamber. The use of over expressed proteins coating the electrode and membrane. In the absence of chemical mediator's, transportation is depended on the mechanical entry and distribution effects of pressure supplied substrate in the blood matrix to facilitate the ECA. Transportation is also depended on effective OCV's and proton concentration gradients. A logarithmic dependence exists where increased enzyme levels exceed the buffer concentrations. Although molecules of similar polarities attract each other the drug permeability will depend on the matrix pH and miscibility of the medium solvents to ensure the diffusing molecule is able to permeate within lipophilic mediums. It seems with time the overall output of the power density decreases however this does not occur in a linear manner. The voltage output was within the range of 0.5 V and 0.8V. This was possible at low glucose concentrations which is promising as OCV increases is then possible at higher glucose concentrations. There was evidence of hydrogen ion consumption (C_o/C_L) ratio increases with the generation and subsequent increased consumption of the proton. This created a voltage gradient also due to the boundary condition near the electrode surface as well as the distance travelled by the proton.

Future micelles with ECA capability for drug delivery following spray drying for microencapsulation, directional flow studies and micro fluidic system design could aid in the effectiveness in prototype advancements to further develop working models. The model will be most effective where the carrier can be inserted into the blood matrix and function optimally utilising the biological conditions without having to make changes to the matrix needed for an optimised OCV.

Acknowledgments
We wish to acknowledge the financial assistance of the National Research Foundation (NRF) and the Stellenbosch University support staff.

References

[1] Atanassov P, Apblett C, Banta S, Brozik S, Barton S C, Cooney M, Liaw B Y, Mukerjee S, and Minteer S D 2007 *Enzymatic Biofuel Cells*. The Electrochemical Society Interface *28-31*
[2] Pathak R and Basu S 2013 Mathematical modelling and experimental verification of direct glucose anion exchange membrane fuel cell *Electrochim. Acta* **113** 42–53,
[3] Basu D and Basu S 2013 Mathematical modelling of overpotentials of direct glucose alkaline fuel cell and experimental validation *J. Solid State Electrochem.* **17** 2927–38,
[4] Wisniewski N, Reichert M 2000 Methods for reducing biosensor membrane biofouling *Coll. Surf. B: Biointerfaces* **18** 197–219

[5] Ganguli R and Dunn B 2012 Electrically conductive, immobilized bio-anodes for microbial fuel cells. *Nanotechnology* **23** 1-7
[6] Naidoo S, Naidoo Q, Blottnitz H V and Vaivars G 2012 Glucose fuelled mediator-less microbial biological fuel cell using crossover limiting nafion membrane at ambient operating conditions https://www.researchgate.net/publication/235420569
[7] Shukla A K, Suresh P, Berchmans S and Rajendran A 2004 Biological fuel cells and their application *Current Science* **87**, 455-68
[8] Erzinger G S, Silveira M M, Vitolo M and Jonas R 1996 Determination of glucose-fructose oxidoreductase activity in whole cells *World Journal of Microbiology and Biotechnology* **1** No. 12 22-4.

IOP Conf. Series: Materials Science and Engineering **503** (2019) 012027 doi:10.1088/1757-899X/503/1/012027

Mass recovery of carbonated fabrics of glass fibres after isothermal heating

E Pentjuss[1], J Balodis[1], M Vdovicenko[2], J Gabrusenoks[1], G Bajars[1], L Petersone[3] and J Kleperis[1]

[1] Institute of Solid State Physics, University of Latvia, Riga, Latvia;
[2] Faculty of physics and mathematics, University of Latvia, Riga, Latvia
[3] Valmieras Stikla Skiedra, AS, Valmiera, Latvia

E-mail: evalds.pentjuss@cfi.lu.lv

Abstract Leaching of Na+ ions in sodium oxide (Na_2O) and silica (SiO_2) containing glass is well investigated mainly due to its weak weathering. The object of this study was naturally (at room conditions) leached, steady state product on surface of sodium oxide-silica-alumina (Al_2O_3) glass fibers (in fabric) in a form of shell of "glyed" trona crystals as a result of interaction of leached Na^+ ions and H_2O and CO_2 from atmosphere. There are presented results of continued former investigation of mass loss by isothermal heating of fabric and mass recovery in different atmospheres during the first phase of adsorption (at least 0.25h) without changes of state of crystals obtained during preheating at different temperatures. There are observed two ways of decomposition of trona ($Na_3H(CO_3)_2{\cdot}2H_2O$) with its beginning at about 55-57^{0}C and 73-75^{0}C. The regression analysis of mass restoring in different atmospheres indicates to simultaneous and exponential mass increase by physical adsorption of CO_2 and H_2O having the different parameters of exponents *vs* time. Decomposition of trona is discussed in terms of parameters of exponent *vs* preheating temperature.

1. Introduction

Glass durability studies are generally focussed on the reaction between water and glass [1]. It has been known that the reaction produces a leached surface layer in which the alkali ions in the glass are depleted to the surface and replaced by hydrogen or hydronium ions from the water to preserve electroneutrality [1-3], that is confirmed experimentally by ion-beam etching and other techniques which determine the concentration profiles of the interdiffusing species as a function of depth in a glass [1,4-7]. The object of this study was the shell of "glued" trona ($Na_3H(CO_3)_2{\cdot}2H_2O$) crystals on surface of sodium oxide-silica-alumina (Al_2O_3) glass fibres (in fabric) as a result of interaction of leached Na^+ ions and surrounding atmosphere during long time (years) storing at room conditions. Another motivation is associated with the small sizes (below a μm in two directions) of crystals that increases the attitude of surface to volume, velocity of reactions, and facilitate the investigation of surface phenomena itself.

The previous investigations [8,9] of mass recovery kinetics of fabric samples showed that during at least the first 0.25 h after heating the mass (or relative mass) recovery follows (R^2>0.99) to regression:

$$A(t) = A_0 - A_1\exp(-t/t_1) - A_2\exp(-t/t_2) = A_1(1-\exp(-t/t_1)) + A_2(1-\exp(-t/t_2)) \quad (1)$$

The coefficients A_1 and A_2 characterize the saturated adsorbed mass volumes, when $t \to \infty$. $A_1+A_2=A_0$. Time constants t_1 and t_2 characterize velocity of mass adsorption. The aim of present work was to elucidate the two mass adsorption components of equation (1) by changing the content of surrounding atmosphere. The second aim was a try to evaluate decomposition of trona at different preheating temperatures by changes of mass adsorption in terms of A_1, t_1 and A_2, t_2 of equation (1). Additionally, there was elucidated the "disparity" of indicated by XRD the one substance of shell (trona) and by differential thermal analysis (DTA) indicated two peaks. It allows evaluate the mass of shell and glass in fabric by TG (thermogravimetry).

2. Experimental

An industrially produced, carbonized and stabilized during years in room atmosphere a glass fabric with composition of (18-22) Na_2O, (3-5) Al_2O_3, (73-79) SiO_2 were used in experiments. The threads contained 600 glass fibres with diameter of 6 μm. To determine composition of fabric, the samples (N15;16) were heated during 2 h at 185^{0}C. To evaluate the content of adsorbed water at room conditions there is used box containing a water absorber. DTA and TG of fabric were fulfilled on Shimadzu DTA-60/60H.

For isothermal experiments there are used a group of samples b, stored in room during winter (less humidity) and samples d stored in room during summer (higher humidity, temperature and its oscillations). Samples b and d were heated from 39.1 to 154.1 or 185 ^{0}C respectively by step of 5^{0}C. Preheating time was 1h. The mass recovery of samples b was determined in three atmospheres: room, box in room at elevated RH about 70%, box in room at elevated RH 70% and CO_2 concentration from 1500 to1700 ppm and samples d: box in room at RH 70%.

The parameters A1, t1 and A2, t2 of equation (1) were determined using a regression technique to recovered mass values of samples for each preheating temperature in dependence on time till 0.25h. Later this time the equation (1) is not valid because there begin the other processes which change the state of crystals [9].

3. Results and discussion

The SEM picture of fibres with shell of trona crystals (XRD) are seen in Figure1. During carbonation the fabric becomes thicker and less flexible, because crystals of shell are "glued" together and with surface of glass. We do not know the possible intermediate reactions on surface and how much time its take. The crystals of shell belong to trona (XRD) that have been converted to soda (Na_2CO_3, XRD) by heating of fabric. The known equation of trona thermal conversion to soda is

$$2Na_2CO_3 \cdot NaHCO_3 \cdot 2H_2O \rightarrow 3Na_2CO_3 + CO_2\uparrow + 5H_2O\uparrow \qquad (2)$$

The curve of DTA signal in Figure 2 during conversion trona to soda of "glued" fabric (in contrary to powder of trona [10]) contains two peaks and corresponding two steps in TG curve (not shown), caused by high thermal conductivity of "glued" system of trona crystals and glass. If "glued" system by pressing the fabric in spherical mortar is broken, the thermal conductivity of system decreases and crystals outward of thread reach the conversion temperature earlier, but inside a thread reach it later, that leads to widening of DTA peak to compare with its "glued" state, as it seen in Figure2. Effect increases as pressing level increases. The decrease of mass at about 40^{0}C in Figure2 could be caused partly by desorption of crystal water of trona (before decomposition of its bicarbonate content [11]).

It is assumed that mass content of fabric before decomposition of trona (P_{bd}) consists of mass of glass (p_{glass}), the steady state mass of adsorbed water at room conditions (p_{water}) and trona crystals:

$$P_{bd} = p_{glass} + p_{water} + p_{trona} \qquad (3)$$

As it proceeds from reaction equation (2), the mass of trona before conversion is

$$p_{trona}= \frac{2M_{trona}}{M_{CO_2}+5M_{H_2O}}\Delta\uparrow=3.3714\cdot\Delta\uparrow \text{ mg} \quad (4)$$

where M_{trona}, M_{CO_2} un M_{H_2O} are molar weights and $\Delta\uparrow$ is loss of mass by heating and evolving of CO_2 and H_2O. Then the composition of fabric is: humidity in range of 0.98-0.99%, trona about 43.8%, leached glass 55.2% from the equilibrium mass of fabric sample at about 22^{0}C and 55% RH.

The relative mass loss (samples b) after/before heating P_1/P_0 of fabric after long and dry storing in room during winter (Figure 3, left curve) indicates the beginning of trona decomposition at about 57^{0}C to compare with 73^{0}C after more humid storing (samples d) in summer, that coincide with observations given for example in [12,13].

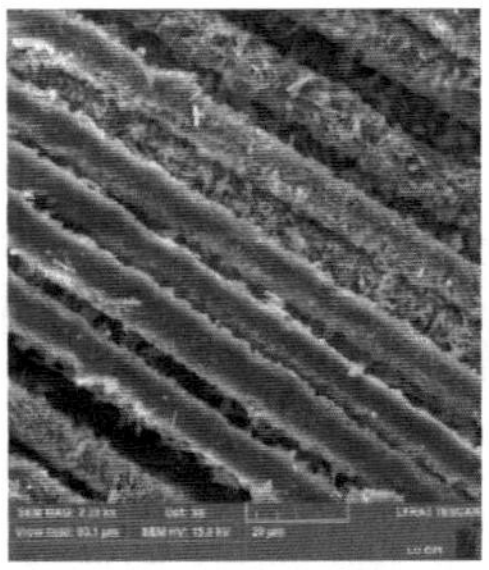

Figure 1. SEM image of shell of trona crystals on fibres of glass

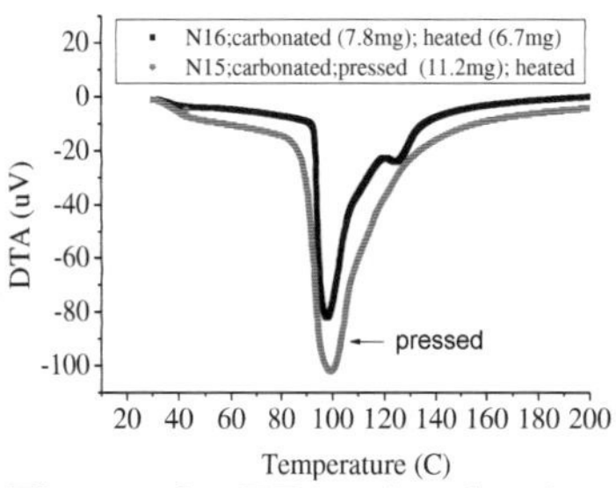

Figure 2. DTA signal of trona decomposition for normal and pressed glass fabrics

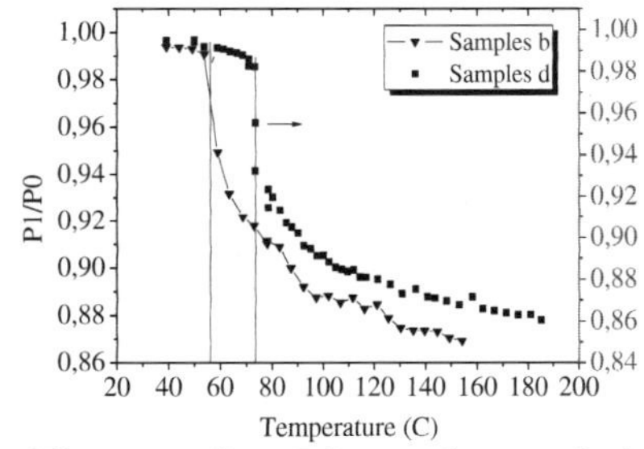

Figure 3. Mass loss during heating(1h) of fabrics stored at dry (left and more humid atmosphere

As it seen in Figure 4, 5 the relation A_1/A_0 increases with concentration of CO_2 and relation A_2/A_0 with humidity of air to compare with its values at normal room conditions. It indicates that the first part of equation (2), having time constant $t_1<<t_2$ (Figure 6) and $A_1/P_0<A_2/P_0$ (Figures 4, 5), characterizes the mass adsorption of CO_2 and the second part- the water adsorption from atmosphere. The mass adsorption of H_2O is higher to compare with CO_2 (Figures 4, 5), but adsorption of CO_2 is faster (Figure 6).

The preheating temperature of the increases of CO_2 and H_2O adsorption in Figure 4, 5 coincides with beginning temperature of trona decomposition in Figure 3 at about 57^{0}C, that could be caused with distortion of surface of crystals by evolving H_2O and CO_2 and increase its area and number of adsorption sites. The dissipation of points in Figure 4-6 is associated with dissipation of temperature and humidity during its measurements. Increase the both of A/P_0 with decrease of temperature (not shown) and increase of concentration of CO_2 and H_2O in atmosphere (Figure 4, 5) indicates to physical adsorption and its simple accumulation during at least 0.25h after those were lost during trona decomposition. The correlation between three mass desorption peaks during decomposition in Figure 3 (left curve) and three peaks of water adsorption (Figure 5) may indicate that adsorption sites are part of the same desorption sites during decomposition, because adsorption is unsaturated and determined by number of adsorption sites that explains also the high selectivity of adsorption.

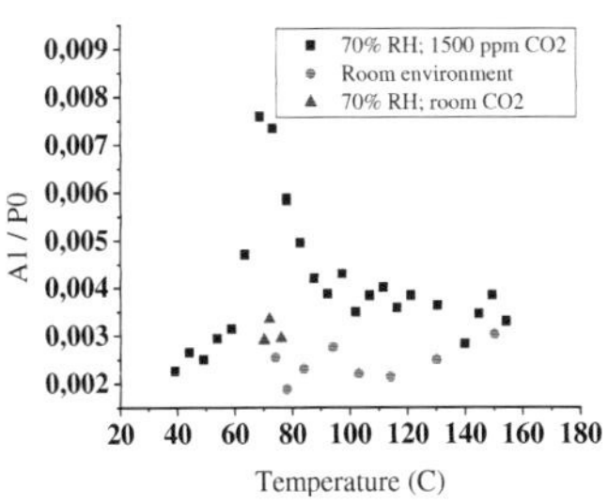

Figure 4. A_1/P_0 at the three atmospheres vs preheating temperature (samples b)

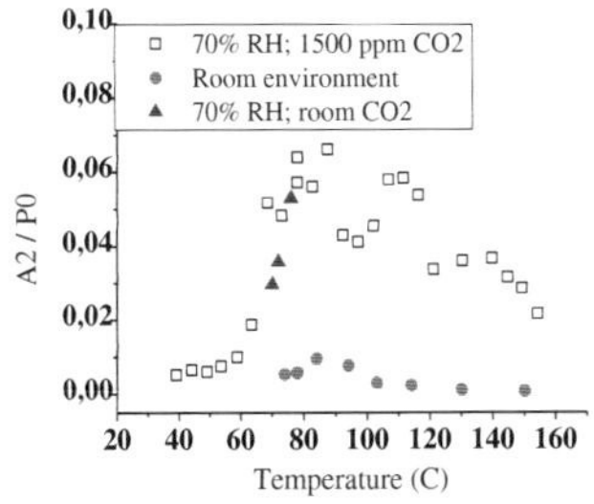

Figure 5. A_2/P_0 at the three atmospheres vs preheating temperature (samples b)

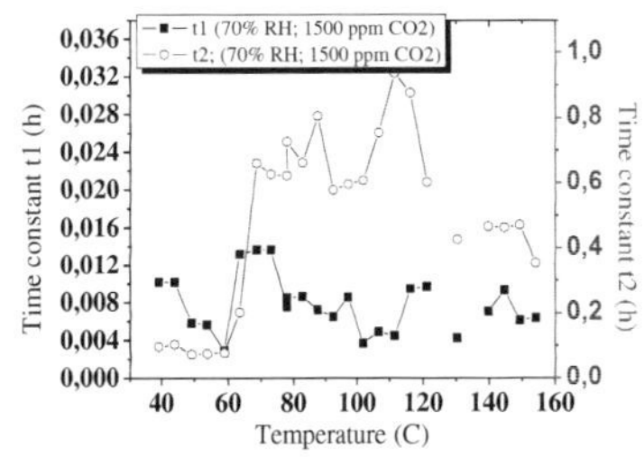

Figure 6. Time constants t_1 and t_2 vs preheating temperature (samples b)

Group of samples d before heating were stored at higher humidity, temperature and they fluctuations (summer) that increase the dissipation of points in Figures 7, 8 compared to that in a case of samples b. Mass after heating was restored at room conditions and increased RH to about 70%. The beginning of decomposition in this case is at 73^{0}C in Figure 3 (right curve), about three times decreased water adsorption compared to samples b (Figure 5), and there is not seen the profiling of curve (Figure 3 right). Adsorption of water A_2/P_0 in Figure 8 also begins at the same temperature as decomposition. Adsorption peak of CO_2 looks similar to that in Figure 4, but wider. The height is the same in spite of lower concentration CO_2 compared to samples b (Figures 4 and 7). It indicates, that crystals of trona remember the earlier storing conditions in character of it decomposition and restoring that are quite different and may contain the new information.

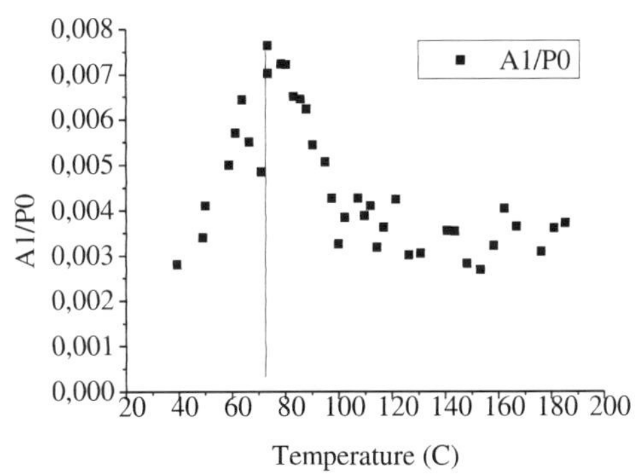

Figure 7. A_1/P_0 vs heating temperature (Samples d, RH 70%)

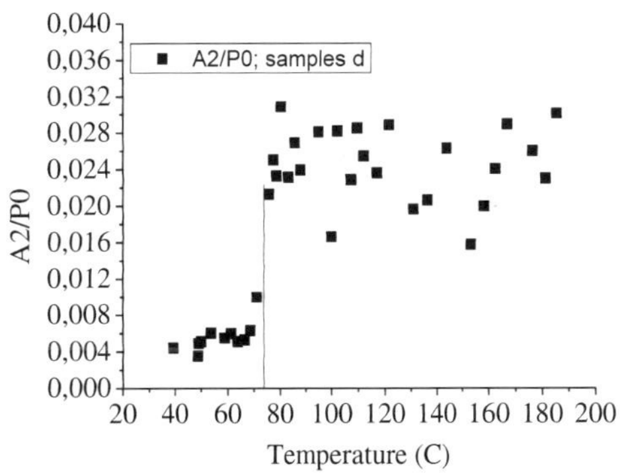

Figure 8. A_2/P_0 vs heating temperature (Samples d, RH70%)

4. Conclusions

The first stage (at least up to 0.25 h) of mass restoring after thermal decomposition of trona crystals is associated with accumulation of mass by physical adsorption of CO_2 and H_2O from atmosphere and approximates with sum of two exponents with different values of its parameters. The first exponent with less time constant of equation (1) characterizes the physical adsorption of CO_2 and the second one - H_2O. Later this stage the mass adsorption moves from exponents.

The increase of fabric adsorption, heated over beginning of decomposition of trona is associated with destruction of surface of crystals, increased adsorption area and number of adsorption sites.

The non-isothermal decomposition of trona in glass fabric has the double and short DTA signal peak, caused by good thermal contact in system of "glued" trona crystals and glass fibre.

Acknowledgement: Authors acknowledge financial support from Latvian National Program IMIS[2]

References

[1] Tsong I S T Houser C A. White W B, Power G L and Tong S S C 1980 *Journal of Non-Crystalline Solids* **38** & **39** 649
[2] Hench L L 1975 *J.Non-Cryst. Solids* **19** 27
[3] Doremus R H 1975 *J. Non-Cryst. Solids* **19** 137
[4] Lanford W A 1978 *Nucl. Instrtum. Meth.* **149** 1
[5] Lanford W A Davis K Lamarche P, Laursen T, Groleau R and Doremus R H 1979 *J. Non-Cryst. Solids* **33** 249
[6] Tsong I S T, Houser C A, Yusef N A, Messier R F, White W B and Michels J W 1978 *Science* **201** 339
[7] Veal B W 1980 *Nuclear Technology* **51** 2 136 DOI: 10.13182/NT80-A32592
[8] Pentjuss E, Lusis A, Bajars G and Gabrusenoks J 2013 *IOP Conf. Series: Materials Science and Engineering*, 49 012044 doi:1088/1757-899X/49/1/012044
[9] Pentjuss E Lusis A, Gabrusenoks J and Bajars G 2015 *IOP Conf. Series: Materials science and engineering*, 77 012021 doi:10.1088/1757-899X/77/1/012021.
[10] Ekmekyapar A, Erşahan H and Yapıcı S 1996(1) *Industrial and Engineering Chemistry Research* **35** 258
[11] Gärtner R S 2005 Mixed solvent reactive recrystallization of sodium carbonate. Proefschrift. Technische Universiteit Delft. ISBN 90-9016165-8; 124
[12] Gärtner R S and Witkamp G-J 2007 *Hydrometallurgy* **88** 75
[13] Copenhaper W C 2009 United States Patent No.:7,638109B2. Date of patent: Dec.29, 2009.

IOP Conf. Series: Materials Science and Engineering **503** (2019) 012028 doi:10.1088/1757-899X/503/1/012028

Characterization of thermoelectric and thermogravimetric properties of conductive PEDOT:PSS films blended with SWCNTs and PVA

G Vugule[1], J Bitenieks[1], J Zicans[1], R Merijs-Meri[1], T Ivanova[1], D Erts[2] and K Buks[2]

[1] Institute of Polymer Materials, Faculty of Materials Science and Applied Chemistry, Riga Technical University, Latvia
[2] Institute of Chemical Physics, University of Latvia, Latvia

e-mail: Janis.Zicans@rtu.lv

Abstract. Poly(3,4-ethylenedioxythiophene):poly(styrenesulfonate) (PEDOT:PSS) was blended with polyvinyl alcohol (PVA) and single-walled carbon nanotube (SWCNT) filler to form composites with thermoelectric properties. Studied samples were obtained by drop coating and solution casting methods. Thermoelectric measurements of PEDOT:PSS demonstrated that the addition of 5 wt. % SWCNTs increased the Seebeck coefficient value from 8.0 μV/K to 23.6 μV/K, while in the case of PEDOT:PSS/PVA blended with 5 wt. % SWCNT Seebeck coefficient value of 20.3 μV/K was achieved. Thermogravimetric analysis showed slight SWCNT effect on thermal stability of the investigated systems.

1. Introduction

In recent years poly(3,4-ethylenedioxythiophene):poly(styrenesulfonate) (PEDOT:PSS) has reached attention as a perspective material for applications in electronics and energetics due to decent thermoelectric properties [1]. Water-soluble PEDOT:PSS is also non-toxic and it is easy to use to obtain uniform conductive films and coatings by means of drop coating and film casting methods that can be applied in laboratory environment [2]. PEDOT:PSS organic nature and thin film properties in future could be applied for commercial manufacturing of small-scale thermoelectric devices for waste energy transformation into electrical power [3].

Addition of conductive nanofillers could be beneficial to further improve PEDOT:PSS thermoelectrical properties. For this purpose single-walled carbon nanotubes (SWCNTs) can be a good candidate due to their wide usage for electrical conductivity enhancement in polymer composites and fact that they can be directly dispersed in PEDOT:PSS without additional surfactants [4]. A good SWCNT dispersion can improve electrical junctions between PEDOT:PSS particles giving contribution to electrical conductivity and thus the thermopower improvement at the same time keeping the increase in thermal conductivity relatively low.

However from PEDOT:PSS composite with carbon nanotubes it is difficult to form freestanding films leaving great negative effect on it flexibility, that may reduce its applicability. Therefore additionally blending of PEDOT:PSS with other polymers offers means to combine the useful and desired properties exhibited by the individual polymer components. In the current case polyvinyl alcohol (PVA) was chosen to modify PEDOT:PSS as it is environmentally friendly, flexible and also water-soluble that can facilitate mixing of these two polymers. Previous studies show that PEDOT:PSS/PVA blends can form free-standing films and reduce the drawbacks of brittleness [5-6] however sacrificing a part of thermoelectric properties due to PVA insulating nature.

2. Experimental

2.1. Materials

PEDOT:PSS aqueous dispersion (Clevios PH 1000) with solids content of 1.3 wt. % was purchased from Heraeus, Germany. SWCNTs with outer diameter 1 – 2 nm, length 5 – 30 μm and purity > 90 % were provided from Timesnano, Chengdu Organic Chemicals Co. Ltd., Chinese Academy of Sciences. PVA (Celvol E 04/88S), 88 % hydrolyzed, was obtained from Celanese, USA.

2.2. Preparation of PEDOT:PSS/PVA/SWCNT films

To prepare PEDOT:PSS/SWCNT composites, SWCNTs were dispersed in PEDOT:PSS solution by ultrasonic probe UIS250V (Hielscher Ultrasonics GmbH). To induce an efficient SWCNT dispersion, sonication time was approximately 20 min. For PEDOT:PSS/PVA/SWCNT sample preparation, first PVA was added to PEDOT:PSS solution at the same concentration (1.3 wt. %) and stirred at 80 °C until the PVA was fully dissolved. Then SWCNTs were added in the same manner as previous. Obtained dispersions were cooled down to room temperature and applied by drop coating method on clean glass substrates for thermoelectric properties characterization. Free standing films were obtained by solution casting on flat polypropylene surface.

2.3. Characterization

The Seebeck coefficient was measured by a custom made device based on the Peltier effect. It consists of two separate Peltier plates which temperature can be adjusted from –20 to +80 °C. Three samples of each PEDOT:PSS nanocomposite system were used. The sample was put into the gap between the Peltier plates to generate temperature gradient. The temperature difference ΔT across the sample was registered with a thermocouple, and the thermovoltage ΔV generated was measured with an HP34401A multimeter. The Seebeck coefficient was determined from the slope of ΔV–ΔT curves. Thermal gravimetric analysis (TGA) according to EN ISO 11358 was carried out using TGA1/SF device (Mettler Toledo Inc.). The change in sample weight was determined at temperature range of 25 – 800 °C at a heating rate of 10 °C/min. Each sample mass was ~ 10 mg and the measurements were performed in nitrogen atmosphere.

3. Results and discussion

3.1. Thermoelectric performance

Measurements of thermopower properties (Figure 1) show that addition of SWCNT filler to PEDOT:PSS up to 1 wt. % results in steep increase in Seebeck coefficient by doubling its value from 8.0 μV/K to 19.6 μV/K. This indicate that dispersed SWCNT filler improves Seebeck coefficient which is governed by formation of electrical percolation network in PEDOT:PSS matrix volume where distance between individual SWCNTs decreases to form electrical contacts between SWCNTs thus improving electrical conductivity of PEDOT:PSS. By further increasing SWCNT concentration up to 5 wt. % the improvement of Seebeck coefficient does not show noteworthy changes. The main reason for such an effect could be that at 1 wt. % SWCNT concentration the composite has reached percolation threshold where additional increase in filler concentration gives no considerable effect.

As expected, addition of PVA, as a typical dielectric polymer, to PEDOT:PSS/SWCNT, reduces the measured thermopower value. However, at higher SWCNT concentrations (> 3 wt. %) it is still possible to achieve some increase in Seebeck effect. The less pronounced enhancement in Seebeck coefficient arises from lower electrical conductivity of the polymer blend matrix due to the fact that PVA hydroxyl groups form hydrogen bonds with PSS sulfonate groups, reducing the negative charge in the PSS [7].

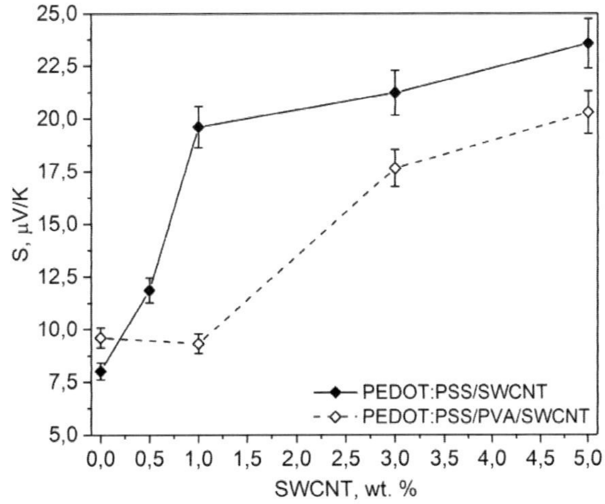

Figure 1. Seebeck coefficient of PEDOT:PSS/SWCNT and PEDOT:PSS/PVA/SWCNT composites.

3.2. Thermal gravimetric analysis

Thermal gravimetric behaviour of the investigated materials is depicted in Figure 2a-b.

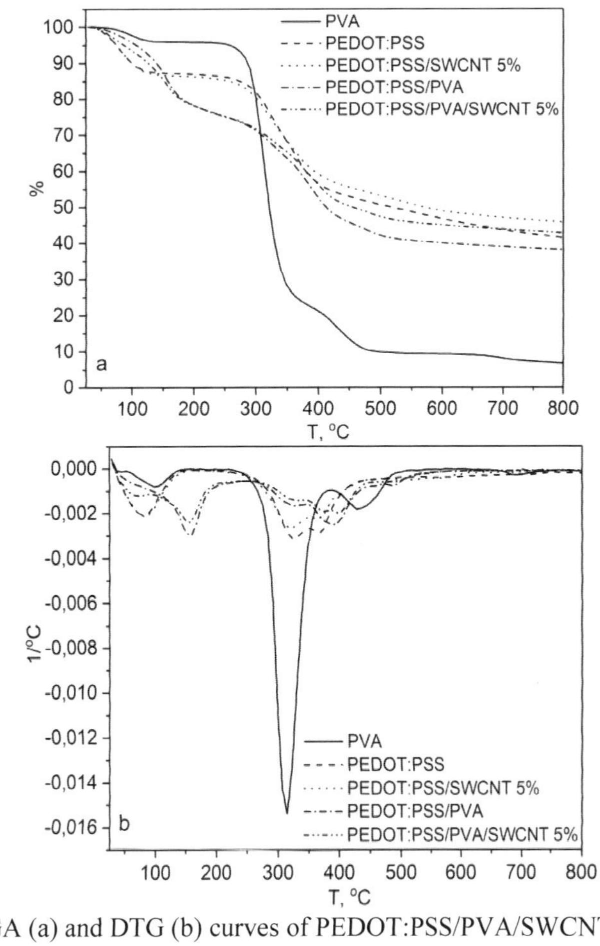

Figure 2. TGA (a) and DTG (b) curves of PEDOT:PSS/PVA/SWCNT composites.

The first weight loss region of PEDOT:PSS sample from 20 to 120 °C is related to release of absorbed water [8], which makes up approximately 12 wt. %. Later, the second weight loss stage, with two overlapping first derivative curve (DTG) peaks, describes decomposition of PSS at 320 °C by rupturing of the sulfonate groups from styrene and PEDOT thiophene chain decomposition at 365 °C [9]. Whereas the first PVA weight loss step with loss of 4 wt. % of absorbed water is located in the temperature range from 50 to 125 °C. Next step is decomposition of hydroxyl groups with peak at 307 °C, followed by smaller weight loss stage at 430 °C (scission of C-C bonds in PVA backbone) [10]. In PVA and PEDOT:PSS blend and its nanocomposite with SWCNT this weight loss step is shifted towards the direction of higher temperatures, most probably due to release of the bound water as a result of hydrogen bonding between PVA hydroxyl groups and PSS sulfonate groups.
In the presence of SWCNT filler residual weight of PEDOT:PSS and PEDOT:PSS/PVA based composite samples is correspondingly increased as well as weight losses of the composites, especially PEDOT:PSS/PVA, up to ca 300°C are promoted. According to thermogravimetric relationships of the investigated composites, it could be estimated that PEDOT:PSS and PEDOT:PSS/PVA based composites could be used for room temperature thermoelectric applications.

4. Conclusions

Thermopower measurement results indicated SWCNT contribution to Seebeck coefficient improvement due to the increase of PEDOT:PSS electrical conductivity which resulted in sharp increase in Seebeck coefficient at 1 wt. % of SWCNT concentration because of formation of electrical percolation network. Blending of PEDOT:PSS with PVA revealed that at SWCNT concentration higher than 3 wt. % it is possible to keep an increase in Seebeck coefficient, even in the presence of insulating polymer matrix however at 14 – 17 % smaller values. TGA measurements of PEDOT:PSS/PVA blends demonstrated that in the presence of SWCNT filler weight losses of the composites, especially PEDOT:PSS/PVA, up to ca 300°C are promoted.

Acknowledgement

Financial support in the framework of the ERDF project Nr. 1.1.1.1/16/A/257 is greatly acknowledged.

References

[1] Li Y, Du Y, Dou Y, Cai K and Xu J 2017 *Synth. Met.* **226** 119
[2] Congcong L, Baoyang L, Jun Y, Jingkun X, Ruirui Y, Zhaojin Z, Shuyun Z, Xiujie H, Zhuo Z and Ping C 2010 *Synth. Met.* **160** 2481
[3] Wei Q, Mukaida M, Kirihara K, Naitoh Y and Ishida T 2015 *Materials* **8** 732
[4] Beesley D J, Price B K, Hunter S, Shaffer M S P and de Mello J C 2016 *Nanocomposites* **2** 135
[5] Mydhili V and Manivannan S 2017 *J. Appl. Polym. Sci.* **134** 45079
[6] Wang Z, Xu J, Yao Y, Zhang L, Wen Y, Song H and Zhu D 2014 *Sens. Actuators* B **196** 357
[7] Chen C, LaRue J C, Nelson R D, Kulinsky L and Madou M J 2012 *J. Appl. Polym. Sci.* **125** 3134
[8] Zhou J, Anjum D H, Chen L, Xu X, Ventura I A, Jiang L and Lubineau G 2014 *J. Mater. Chem.* C **2** 9903
[9] Antiohos D, Folkes G, Sherrell P, Ashraf S, Wallace G G, Aitchison P, Harris A T, Chen J and Minett A I 2011 *J. Mater. Chem.* **21** 15987
[10] Holland B J and Hay J N 2001 *Polymer* **42** 6775

IOP Conf. Series: Materials Science and Engineering **503** (2019) 012029 doi:10.1088/1757-899X/503/1/012029

Model colloids to study surface - ligand interactions in nanosized Fe_3O_4

M M Maiorov[1], D Zablotsky[1], E Blums[1] and A Krumina[2]

[1] Institute of Physics, University of Latvia, Latvia
[2] Institute of Inorganic Chemistry, Riga Technical University, Latvia

e-mail: maiorov@sal.lv

Abstract. Using electrodispersion of metallic iron in an aqueous environment without additional reagents we create a chemically clean model system based on nanosized synthetic magnetite where the details of magnetic metal oxide - surfactant chemisorption can be studied by high fidelity magnetometry with nanometric precision. A comparative analysis of model colloidal solutions based on oleic acid and iron Fe^{2+} oleate indicates a significant influence of the mode of surfactant adsorption on the magnetic diameter of the particles.

1. Introduction

Solubilized magnetic nanoparticles sized in the low nano-range (below 10 nm) are prospective vehicles for guided drug transport (magnetic drug targeting) [1], potential therapeutic means enabling magnetic hyperthermia [2] and candidates for next generation T2*/T2 contrast agents achieving negative contrast enhancement in magnetic resonance imaging, several of which were approved for clinical use (Ferumoxsil, Ferumoxide, Ferucarbotran etc.) and some in different phases of clinical trials [3][4]. Relative chemical inertness and biocompatibility [5] are required for these purposes and superparamagnetic iron oxide based nanoparticles (SPIONs), which are predominantly produced by chemical condensation of the ferromagnetic phase (co-precipitation, hydrothermal synthesis, thermolysis), best fulfil this role [1][6][7]. However, the colloids obtained from chemical synthesis are always polluted with residual by-products requiring many purification steps before they could be used in biomedical applications.

In turn, surfactants play a vital role to enable solubilisation and solution processing of nanoparticles via a simple, easily scalable and highly versatile colloidal route. Among those, carboxylic acids (e.g. oleic acid) are used extensively in the chemical synthesis and stabilization of nanomaterials based on metal oxides. However, the difficulty of exercising precise compositional control over the solutions produced by a chemical route obscures many aspects of the surface - ligand interactions. The oleic acid (OA), which is by far the most employed surfactant to effectuate colloidal systems based on magnetic iron oxide (magnetite), binds tightly to the surface of nanoparticles with its carboxylate group forming an adsorption layer that prevents the coagulation of the particles. Meanwhile, this surfactant acting as an acid may chemically interact with the metal forming non-ferromagnetic salts – oleates. This behaviour may eventually affect the disperse composition of the formed colloidal system and is undesirable.

Herein, we present a comparative analysis of two model colloidal systems, produced by an alternative non-chemical route, one stabilized by oleic acid [8][9] and the other by iron Fe^{2+} oleate salt

(Fe^{2+}OL), which acts as a surfactant, but where carboxylate groups are passivated by bidentate coordination to a Fe^{2+} ion [10]. We prepare a clean reference system by physical condensation that is based on electric ablation of metallic granules immersed in a dielectric fluid by high frequency-pulsed spark discharges forming a luminous network of conductive channels between the granules in the particle bed and the electrodes [11]. The vaporized material is ejected at high temperature and rapidly quenched *in situ* condensing in the form of nanoparticles with well-defined spherical morphology. Electric dispersion uses no chemicals except the original substance from which the nanoparticles are produced, utilizes only electric power and does not require chemical precursors or produce waste, hence, avoiding the ambiguity of chemical composition characteristic for commonly used synthetic methods.

2. Experimental section

Synthesis of nanoparticles. A 10 g portion of biocompatible coarse-grained metallic iron powder (pharmaceutical iron *Ferrum reductum*, intended for therapeutic oral administration) with particles sized 0.3-1.5 mm (mesh -14+50) of no particular shape was added to 200 ml of distilled water. Flat steel electrodes connected to a spectrometric discharge generator operating at 100 Hz were pressed onto the particle bed. After turning on the generator a luminous network of sparks forms within the particle bed. Reactive dispersion of iron using distilled water as the working dielectric is accompanied by rapid oxidation of ejected iron and the formation of magnetite [11]. After 15 min of operation the initially transparent liquid turns completely black evidencing the condensation of nanodispersed phase. Then the supernatant is decanted and centrifuged, after that the precipitate is washed several times with acetone and dried. The distinct advantage of producing magnetite nanoparticles by electrodispersion of iron in distilled water is that the absence of chemical by-products, such as salts, acids or bases, which accompany chemical synthesis, in the obtained materials is assured.

Synthesis of iron Fe^{2+} oleate. The salt is prepared by a reaction of metal chloride and sodium oleate [12]. For this purpose 3.6 g portion of iron(II) chloride tetrahydrate ($FeCl_2 \cdot 4H_2O$, Sigma-Aldrich, 99%) was dissolved in 12 ml of distilled water and 14 g of sodium oleate (Sigma-Aldrich, 82 %) was dissolved in a mixture comprising 6 ml of distilled water, 24 ml of ethanol and 42 ml of *n*-hexane. The resulting solutions were transferred to a reactor heated to 70 °C and mixed for 4 h under argon flow. The mixture was transferred to a separation funnel and the upper organic layer containing hexane solution of iron Fe^{2+} oleate was decanted and washed several times with distilled water.

Synthesis of model colloids (Fe_3O_4)@OA and (Fe_3O_4)@Fe^{2+}OL. To establish a reliable reference the protocol for the synthesis of model colloidal systems is the same, only changing the type of used surfactant: for model system (Fe_3O_4)@OA the oleic acid (Sigma-Aldrich, 90%) was used, whereas for model system (Fe_3O_4)@Fe^{2+}OL the iron Fe^{2+} oleate synthesized in the previous step was employed. Initially, series of surfactant solutions were prepared at various concentrations of employed surfactant in undecane – an alkane hydrocarbon solvent, which has a convenient temperature range and well dissolves OA. A portion of 0.3 g of dried powder of nanoparticles produced by electrodispersion was added to 10 ml of surfactant solution and redispersed by treating with ultrasound for 30 min at nominal output power of 30 W. Subsequently the suspension was centrifuged for 10 min at 4000 g. The supernatant containing a stable colloidal solution was decanted and retained for further characterization.

Characterization of materials. The model colloidal solutions (Fe_3O_4)@OA and (Fe_3O_4)@Fe^{2+}OL produced at various concentrations of surfactant solutions were studied by vibrating sample magnetometry (vibrating sample magnetometer Lake Shore Cryotronics, Inc., model 7404VSM) - a non-invasive and non-destructive investigative technique for probing the susceptibility, magnetization or hysteresis of magnetic materials that can provide information on the nanoscopic scale. Briefly, a small sample (~40 mg) of liquid material is magnetized within a region of uniform magnetic-flux density with variable strength (-10 kOe...+10 kOe) and subjected to vibrational excitation at low frequency, and the magnetic moment of the sample is deduced.

3. Results and Discussion

A characteristic magnetization curve measured for a sample based on $(Fe_3O_4)@Fe^{2+}OL$ model system (Figure 1, left) shows a superparamagnetic response with no hysteresis, which is modulated by a paramagnetic contribution due to the presence of a paramagnetic salt – iron oleate – in the solution. The correction of the magnetization against parasitic processes in the first order of magnetic field H is straightforward and can be achieved without difficulty

$$\sigma^*(H) = \sigma(H) - \gamma H$$

where γ is the effective paramagnetic susceptibility.

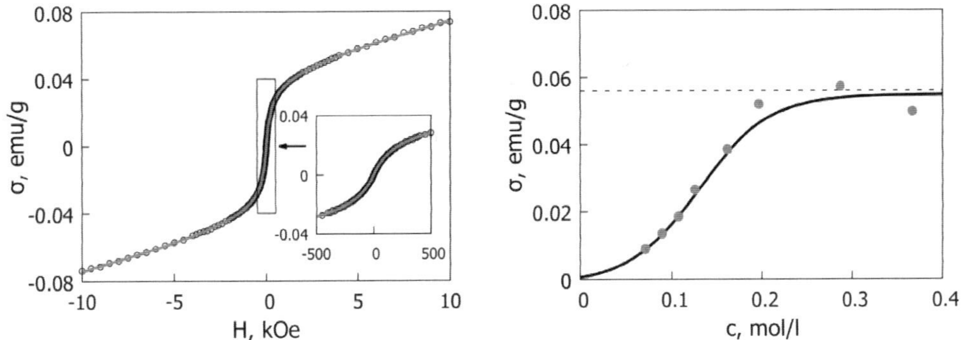

Figure 1 Results of magnetometric measurements of a model colloid based on $(Fe_3O_4)@Fe^{2+}OL$: left – measured magnetization curve, inset shows detailed view of the initial region; right – saturation magnetization as a function of surfactant content in the solution.

The saturation magnetization $\sigma_S = \sigma^*(H \to \infty)$ shows a dependence on the surfactant content in the solution (Figure 1, right), which is initially a power law at low iron Fe^{2+} oleate concentration and gradually reaches saturation. The method of determining the size distribution of magnetic nanoparticles from the magnetization curve is based on representing the corrected curve $\sigma^*(H)$ in the form of a functional, whose kernel is the Langevin function $L(x) = \coth x - \frac{1}{x}$. In CGS system:

$$\sigma^*(H) = I_s \int_{m_1}^{m_2} f(m) L\left(\frac{mH}{kT}\right) dm$$

where $f(m)$ is the volume density of the distribution function of magnetic moments m of the particles and I_s is the spontaneous magnetization of the particle material. The reconstruction is quite accurate as evidenced by the solid line in Figure 1, left.

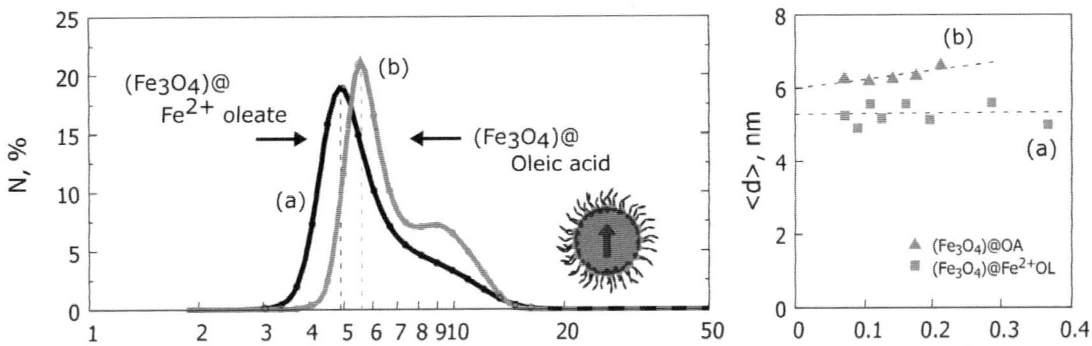

Figure 2 Results of magnetometric analysis: left - particle size distribution of model colloidal system (a) $(Fe_3O_4)@Fe^{2+}OL$ based on iron Fe^{2+} oleate, (b) $(Fe_3O_4)@OA$ based on oleic acid; right – corresponding dependence of average magnetometric size on surfactant concentration.

In the magnetogranulometric procedure the obtained distribution of magnetic moments unambiguously corresponds to the particle size distribution, because the magnetic moment of a superparamagnetic particle is proportional to its volume [13]. Our results indicate that the surfactant content does not appreciably affect the reconstructed size distribution. A typical example for both model systems is shown in Figure 2, left and has a slightly bidisperse structure. However, the average magnetic size $\langle d \rangle$ in the case of oleic acid based colloid (Fe_3O_4)@OA is noticeably larger than the system based on iron Fe^{2+} oleate (Fe_3O_4)@Fe^{2+}OL. This discrepancy is robustly observed independently of the surfactant loading in the solution (Figure 2, right).

4. Conclusions

Here we have systematically studied two model colloids based on a well defined reference system produced by a non-chemical route – physical condensation via electrodispersion, which uses no chemicals except the original substance from which the nanoparticles are produced, without any additional precursors – one stabilized by a covalently bound surfactant and the other with non-covalent binding. It appears that due to their high surface energy the OA either completely dissolves the ultrasmall particles (below approx. 4-4.5 nm) and converts the contained material into an oleate salt or somehow affects the magnetically "dead" layer on the surface of these particles making them invisible to our analysis. The prepared model systems present a unique opportunity to study size focusing regimes [14] in thermolytic processing of oxide nanocrystals with nanometric precision.

Acknowledgments

This work was supported by the SRP IMIS[2], DZ acknowledges the support of M-ERA.NET project No. 3184 and the postdoctoral research program at the University of Latvia (Project No. 1.1.1.2/VIAA/1/16/072).

References

[1] Zablotskaya A, Segal I, Maiorov M, Zablotsky D and Blums E 2018 Iron oxide/oleic acid magnetic nanoparticles possessing biologically active choline derivatives *Inorganic Frameworks as Smart Nanomedicines* ed A M Grumezescu (Elsevier) chapter 7 pp 279–316
[2] Maiorov M M, Blums E and Kronkalns G 2008 *Magnetohydrodynamics* **44** (1) 27–32
[3] Bao G, Mitragotri S and Tong S 2013 *Annu. Rev. Biomed. Eng.* **15** 253-82
[4] Wang YX 2015 *World J. Gastroenterol.* **21** 13400-02.
[5] Zablotskaya A, Segal I, Maiorov M, Zablotsky D, Mishnev A, Lukevics E, Shestakova I and Domracheva I 2007 *JMMM* **311** 135–9
[6] Tartaj P, Morales M P, Veintemillas-Verdaguer S, Gonzalez-Carreno T and Serna C J 2003 *J. Phys. D: Appl. Phys.* **36** R182–97
[7] Mahmoudi M, Sant S, Wang B, Laurent S and Sen T 2011 *Adv. Drug Deliv. Rev.* **63** 24–46
[8] Mozgovoi E N and Blum E 1971 *Magnetohydrodynamics* **7** 448–53
[9] Berkowitz A E and Walter J L 1983 *JMMM* **39** 75–8
[10] Bronstein L M, Huang X, Retrum J, Schmuker A, Pink M, Stein B D and Dragnea B 2007 *Chem. Mater.* **19** 3624–32
[11] Maiorov M, Blums E, Kronkalns G, Krumina A and Lubane M 2016 *Magnetohydrodynamics* **52** 507–12
[12] Park J, Kwangjin An, Hwang Y, Park J-G, Noh H-J, Kim J-Y, Park J-H, Hwang N-M and Hyeon T 2004 *Nat. Mater.* **3** 891–5
[13] Maiorov M M, Blums E and Raj K 2010 *Physics Procedia* **9** 74–7
[14] Tiju T, Sivaramakrishnan S, Divy S, Dinesh K, Bakthavachalam K, Amal A, Sai P, Rogith V, Saif K, Pranthosh D, Manish K, Kedarnath B, Yeshwanth C and Bhusankar T 2018 *Appl. Surf. Sci.* **448** 248–53

IOP Conf. Series: Materials Science and Engineering **503** (2019) 012030 doi:10.1088/1757-899X/503/1/012030

Different approaches in sulfonated poly (ether ether ketone) conductivity measurements

D Fedorenko[1,2] and G Vaivars[1,2]

[1]Faculty of Chemistry, University of Latvia, Jelgavas Street 1, Riga, LV-1004, Latvia
[2]Institute of Solid State Physics, University of Latvia, Kengaraga Street 8, Riga, LV-1063, Latvia

E-mail: deniss.fedorenko@gmail.com

Abstract. Ion conductivity of sulfonated poly (ether ether ketone) (SPEEK) membranes with various degree of sulfonation (DS) was investigated using impedance analysis with different measuring cell configuration and ion conductivity was calculated from resistances of polymer membranes. SPEEK was synthesized from poly (ether ether ketone) (PEEK) via sulfonation reaction in concentrated sulfuric acid (95–98%). Scanning electron microscopy (SEM) analysis of membrane surface was performed to determine possible mechanical damage to the membrane during resistance measurements.

1. Introduction

In modern society there is increasing interest towards innovative technology as well as transition to sustainable energy. Polymer electrolyte membrane is a key component in fuel cell technology, as it provides ion transport in fuel cell [1]. Sulfonated poly (ether ether ketone) (SPEEK) is promising material for application in direct methanol fuel cell [2]. And it was reported that the SPEEK membrane might be cross-linked to enhance its properties [3]. Ion conductivity of polymer electrolyte membrane is a key parameter for providing high efficiency and therefore correct conductivity measurements are very important. Impedance analysis is being used widely to determine membrane conductivity. The literature data on specific cell configuration is limited. At the same time the methods for measuring conductivity of polymer membrane were not described sufficiently by authors previously. The membrane contact with the electrode is important and the contact resistance is difficult to evaluate. That's why it is considered as a probable reason for decreased conductivity. It was also mentioned, that the conductivity of a membrane can be measured by two probe method and slight anisotropy of proton conductivity is observed and through-plane setup showed slightly higher proton conductivity than in-plane setup. Four probe method is limited for polymer membranes due to the low material hardness and strong dependence on relative humidity [4].

The aim of our study is to compare two different methods of measuring ion conductivity of polymer electrolyte membranes using impedance analysis and evaluating them as not being damaging for membranes. First, the standard method by pressing sample in-between two metal electrodes was used. Second, the membrane was pressed in-between two Nafion membranes with a known conductivity (differential method). In this paper two approaches of ion conductivity measurements have been applied for SPEEK membranes with different degrees of sulfonation (DS) using resistance data obtained by

IOP Conf. Series: Materials Science and Engineering **503** (2019) 012030 doi:10.1088/1757-899X/503/1/012030

performing impedance analysis. It was revealed, that the differential method as compared to the single membrane method shows good correlation with a reference material, which was commercial Nafion N-117 membrane.

2. Experimental

SPEEK was synthesized from poly (ether ether ketone) (PEEK) via sulfonation with concentrated sulfuric acid (95–98%) at 60°C. Degree of sulfonation was determined using titration method as described previously [3, 5]. Metrohm Autolab potenciostat/galvanostat with FRA was used to determine the membrane resistance in a frequency range from 100 Hz to 50 kHz and signal amplitude 10 mV. Measurements were taken at 22°C temperature. The membranes were immersed in a deionized water for 24 h and maintained RH = 100% in a measuring chamber.

Conductivity measurements of SPEEK membranes as well as Nafion membranes were performed through-plane using impedance analysis with various cell configurations. In differential method, the SPEEK membrane was sandwiched between two Nafion membranes and pressed between two copper electrodes (1 cm in diameter). The resistance R_1 was obtained from Nyquist plot extrapolating to the high frequencies. Using the same method, resistance of two Nafion membranes (R_2) was determined and as a result, from the difference between two of these measurements, resistance $R_{membrane}$ and conductivity of SPEEK membrane was calculated using equation (1).

$$R_{membrane} = R_1 - R_2 \qquad (1)$$

In case of the single membrane method the SPEEK membranes were pressed between two copper electrodes. Impedance analysis was performed, and the resistance was found from Nyquist plot. The Nafion membrane was used as a reference. The conductivity of polymer membrane using differential and single membrane configuration was calculated from the complex resistance data from Nyquist plot and the results were compared with the literature data.

Scanning electron microscopy (SEM) method was used to inspect the membrane surface and characterize the mechanical damage before and after conductivity measurements for both SPEEK and Nafion membranes.

3. Results and discussion

In our work two methods for measuring conductivity are assessed. Results were compared with a literature data. In both cases, the same membrane preparation method was used and *N,N*-dimethylformamide as a solvent. Membrane thickness ranged from 0.12 to 0.18 μm and was measured with digital micrometer. Electrode surface area was 0.785 cm^2. Figure 1 shows cell configuration for differential and single membrane method. The variation of data (Figure 2, Table 1 and 2) is significant.

The Nafion and SPEEK membrane conductivities that were obtained by using both methods are shown in Table 3. The Nafion conductivity according to the product information is 0.10 S/cm and the same value was reproduced using differential method. As we can see from the Figure 2, the SPEEK membrane conductivity as a function of DS might be plotted as a line, if the same method is used for measurements. However, the data from both methods vary significantly. We can conclude that the contact resistance between electrodes and membrane is quite high. The usage of differential method allows to exclude it efficiently. From such point of view evaluating the literature data, it is evident that the contact resistance was the reason also for high variation of literature data.

It is worth mentioning that the cell stability can be evaluated via Kramer–Kronig test and during our measurements, the function χ^2, χ^2_{re} and χ^2_{im} values were between 10^{-5} and 10^{-6} and that means that the cell was stable, and equilibrium was not disturbed during impedance analysis.

Figure 3 shows an example of data for SPEEK membrane with DS = 0.82 impedance analysis with differential method. Nyquist and Bode plot were presented for cell configuration that consists of two Nafion membrane and one SPEEK membrane between them. Resistance is obtained from extrapolation to the high frequencies where imaginary impedance is equal to zero.

Extrapolation with semicircle method cannot be used in this case because full semicircle is not obtainable and therefore linear regression method is being used. Bode plot shows that the contact resistance is significant. Consequently, this explains higher resistance and lower conductivity for single membrane method and it shows that the differential method excludes contact resistance more efficiently.

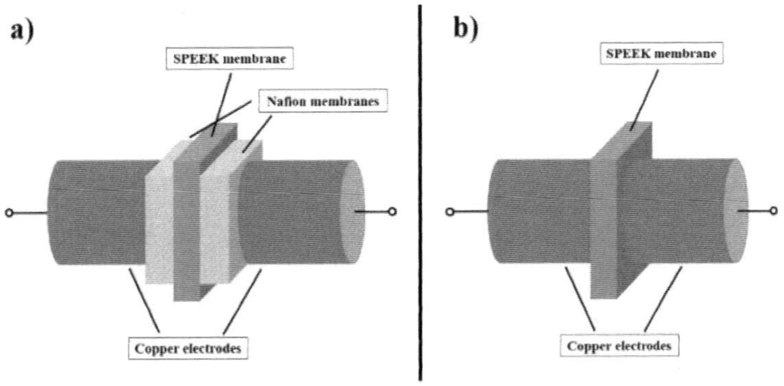

Figure 1. Representation of cell configuration: a) differential method, b) single membrane method

Table 1. Ion conductivities of SPEEK membranes as measured by differential method.

DS	σ. S/cm
0.34	0.0036
0.57	0.0071
0.71	0.0112
0.82	0.0146
0.89	0.0156

Table 2. The differential method as compared with a single membrane method.

Method	Ion conductivity σ, S/cm	
	SPEEK (DS = 0.82)	Nafion
Differential	0.015	0.095
Single membrane	0.0028	0.04

Table 3. Ion conductivity of SPEEK membranes from literature data.

DS	σ, S/cm	Reference	DS	σ, S/cm	Reference	DS	σ, S/cm	Reference
0.30	0.0060	[6]	0.67	0.0022	[8]	0.80	0.0114	[9]
0.36	0.0015	[7]	0.67	0.0075	[9]	0.81	0.0250	[6]
0.48	0.0020	[8]	0.69	0.0038	[10]	0.83	0.0072	[7]
0.48	0.0025	[7]	0.70	0.0013	[11]	0.84	0.0123	[9]
0.57	0.0039	[7]	0.74	0.0021	[11]	0.87	0.0134	[9]
0.57	0.0042	[9]	0.74	0.0091	[9]	0.88	0.0033	[10]
0.59	0.0009	[10]	0.78	0.0130	[6]	0.90	0.0095	[7]
0.63	0.0090	[6]	0.79	0.0160	[8]			
0.65	0.0059	[7]	0.80	0.0180	[6]			
0.67	0.0020	[10]	0.80	0.0080	[11]			

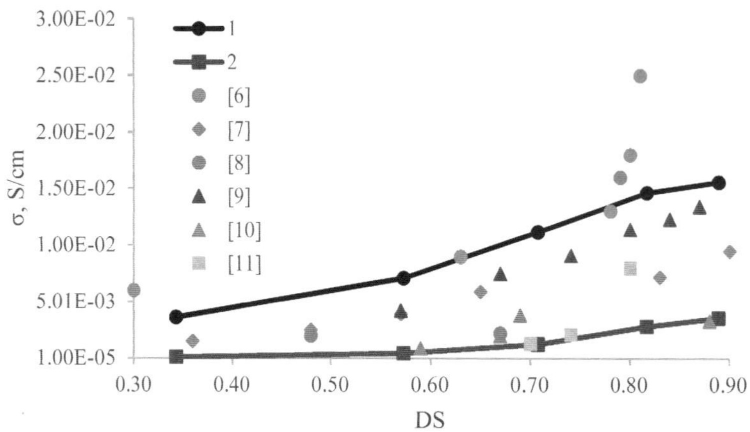

Figure 2. Ion conductivities of SPEEK membrane from literature data and by using two measuring methods: differential (1) and single membrane (2) method.

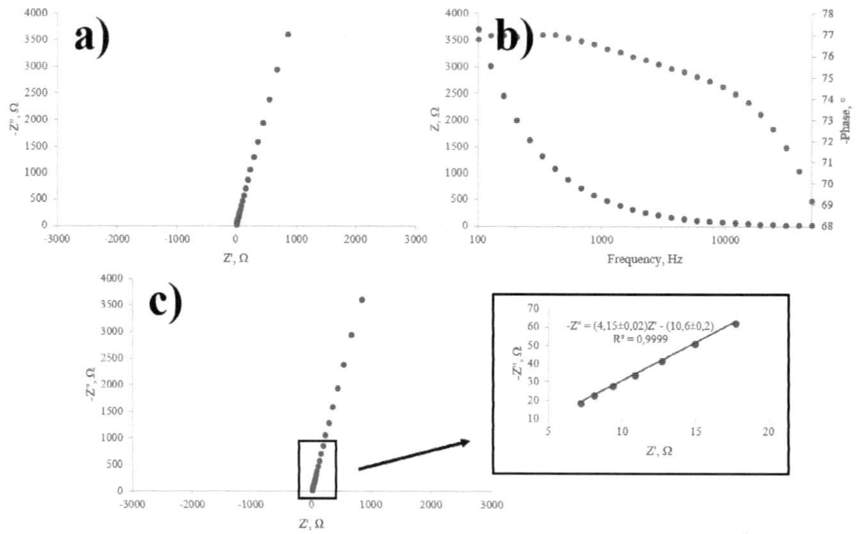

Figure 3. Impedance analysis for SPEEK membrane with a DS = 0.82 by applying differential method with two Nafion membranes: a) Nyquist plot, b) Bode plot, c) obtaining the resistance from Nyquist plot by extrapolating to the high frequencies.

SEM analysis was used to inspect the membrane surfaces. As an example, the SPEEK membrane with DS = 0.82 is presented. Figure 4 reveals some characteristic surface defects, which were produced during synthesis process. After impedance analysis with a differential method no additional defects were observed, but after measuring with a single membrane method, it is obvious, that there are significant scratches. Similar pattern was observed by inspecting Nafion membranes. After impedance analysis the increasing mechanical damage of the Nafion membrane surface could be observed. SPEEK membranes are less elastic, so the mechanical damage is also more pronounced.

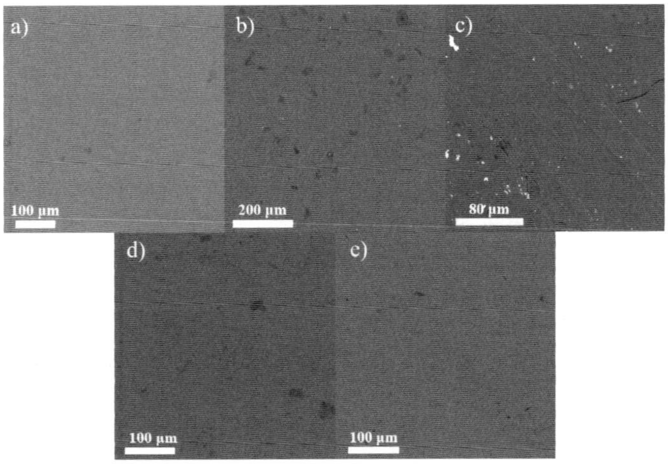

Figure 4. Scanning electron microscope images of SPEEK membrane with DS = 0.82: a) before impedance analysis, b) after impedance analysis with a differential method, c) after impedance analysis with a single membrane method, d) Nafion membrane before impedance analysis and e) Nafion membrane after impedance analysis with a differential method.

4. Conclusion

In this study sulfonated poly (ether ether ketone) membranes were studied using impedance analysis. Differential method proved to be more accurate in conductivity measurements as compared to the single membrane method. Commercial Nafion membrane was used as a reference. SEM analysis revealed also that the differential method is less damaging to the membrane surface. Therefore, the differential method in impedance analysis proved to be more efficient. The membranes are typically used in different solid state ionic devices and the same membrane might be used in a device also after conductivity measurements. The variety of distribution of literature data might be explained by variety of the contact resistance between membrane and measuring electrodes.

Acknowledgement

Authors of this work acknowledge funding from European Union`s Horizon 2020 Research and Innovation Program project under grant agreement No 768789.

References

[1] Hinds G 2017 *Curr. Opin. Electrochem.* **5** 11–9
[2] Luo H, Vaivars G and Mathe M 2010 *J. Power Sources* **195** 5197–200
[3] Luo H, Vaivars G and Mathe M 2012 *Int. J. Hydrogen Energy* **37** 6148–52
[4] Soboleva T, Xie Z, Shi Z, Tsang E, Navessin E and Holdcroft S 2008 *J. Electroanal. Chem.* **622** 145–52
[5] Boaretti C, Roso M, Lorenzetti A and Modesti M 2015 *Material* **8** 4096–117
[6] Luo H Proton conducting polymer composite membrane development for Direct Methanol Fuel Cell applications PhD Thesis 2008 University of the Western Cape Cape Town
[7] Yu F and Hu J 2016 *Int. J. Electrochem. Sci.* **11** 724–37
[8] Xing P, Robertson G P, Guiver M D, Mikhailenko S D, Wang K and Kaliaguine S 2004 *J. Membr. Sci.* **229** 95–106
[9] Xi J, Li Z, Yu L, Yin B, Wang L, Liu L, Qiu X and Chen L 2015 *J. Power Sources* **285** 195–204
[10] Kaliaguine S, Mikhailenko S D, Wang K P, Xing P, Robertson G and Guiver M 2003 *Catal. Today* **82** 213–22
[11] Zaidi S M J, Mikhailenko S D, Robertson G, Guiver M and Kaliaguine S 2000 *J. Membr. Sci.* **173** 17–34

IOP Conf. Series: Materials Science and Engineering **503** (2019) 012031 doi:10.1088/1757-899X/503/1/012031

Nafion® and polyaniline composite modification with Li and Mg ions

Stanislavs Lozkins[1, 2], Julija Hodakovska[1], Janis Kleperis[1] and Remo Merijs-Meri[2]

1 - Institute of Solid State Physics, University of Latvia, Latvia

2- Department of Polymer Materials Technology, Riga Technical University, Latvia

stanislavski@inbox.lv

Abstract.
Modification of Nafion®-based samples with polyaniline (PANI) and positive charged ions could potentially allow achieving good proton conductivity at higher working temperatures what is observed in typical polymer electrolyte membrane fuel cells.

In this work Nafion® polymer was modified with 1% polyaniline emeraldine salt and Li and Mg ions containing solutions. The influence of activation process and the type of ions on composites' proton conductivity was investigated. To define the best composition the physiochemical properties of synthesized materials were determined by the means of thermogravimetric analysis, impedance spectroscopy, determination of water absorption and its kinetics, and FTIR. From the original samples the best results of σ=155 mS/cm were shown by Nafion®/PANI membrane what was modified by Li+ reduction solution. This membrane also had better thermal characteristic compared to other researched samples: membrane degradation starts on higher temperature interval.

1. Introduction

Nowadays commercially available for applications in fuel cells are perfluorinated polymers, such as Nafion® made by DuPont Company. Many other materials were synthesized and tested, but till now Nafion® has shown better properties' relation. Main advantages of Nafion® are chemical and thermal stability, high proton conductivity, also mechanical stability. However, membrane's practical usage has been limited by Nafion® membrane's disadvantages, e.g. low proton conductivity at low humidity, high fuel (hydrogen or methanol) cross-over and finally high price of Nafion® [1]. The ways to eliminate this disadvantage is to synthesize new materials or modify existing ones, both could be done with a set of methods and could bring desirable improvements of exploitation values.

The focus of this paper is to synthesize and study the physical properties of Nafion®/PANI composite membranes, such as the water uptake, degradation and proton conductivity. Nafion®/PANI membrane was prepared by mixing Nafion® and PANI solutions and casting on glass and then treated by lithium and magnesium ions containing solution. Permeability and ionic conductivity of the modified membranes were measured using electrochemical impedance spectroscopy; the thermal stability was determined by thermogravimetric analysis, and the presence of ions in polymer matrix was determined with FTIR spectroscopy. The conductivity and thermal stability of all samples were compared with the unmodified Nafion® membrane as a reference.

2. Experimental

2.1. Materials

N-Methyl-2-pyrrolidone (NMP), 96% sulphuric acid (H_2SO_4), polyaniline emeraldine salt powder were supplied by Sigma-Aldrich. Nafion® (DuPont, equivalent weight of 1100 g/mol) was purchased as 10 w% water solution from Fuel Cell Store. All chemicals were used without further purification. Li^+ and Mg^+ ions containing reducing solutions were prepared from lithium or magnesium hydroxides accordingly by adding concentrated hydrogen chloride to get Li or Mg chloride water solution.

2.2. Membrane preparation

To prepare composite membrane Nafion® solution has been dried in air, than both polymers were separately dissolved in NMP and stirred for 24 h to obtain homogeneous solution. Solutions were mixed together to achieve polymer relation of Nafion®:PANI=99:1 and stirred for 24 h until homogeneous solution was obtained. The solution was cast on a glass plate, which was kept at temperature of 60°C, and after solvent evaporation a membrane was obtained. The membrane was washed in deionized water to remove residuals of solvent; membranes were activated by 4 steps: 1) samples were treated 1 hour by hydrogen peroxide on 80ºC; 2) samples were treated 30 minutes by distilled water on 80ºC; 3) samples were treated by sulfuric acid (96%) on 80ºC; 4) membranes were washed in deionized water on room temperature. Then membrane was processed by reducing solution.

2.3. Methods

2.3.1. Impedance spectroscopy. The proton conductivities were determined using AC impedance spectroscopy with Becktech conductivity cell connected to potentiostat VoltaLab PGZ301 and self-made humidifying system. The samples in the cell were heated up to 70 °C with 100% humidity; frequency range was 2 mHz – 100 kHz. The conductivity values of the prepared membranes were calculated using the following equation: $\sigma=1/\rho$, where ρ - electrical resistivity, calculated from impedance spectrum.

2.3.2. Differential thermal analysis. Thermogravimetric analysis was performed with thermogravimetric analyser Shimadzu's DTG-60 in the region between 30 °C and 600 °C with heating speed of 10 °C/min in argon flow 20 ml/min.

2.3.3. Water uptake. The membranes were cut into samples with equal area, dried, weighted and then placed in deionized water at room temperature for a chosen time. The water uptake values of the prepared membranes were calculated using the following equation: $Water\ uptake\ (\%) = \frac{w_s - w_d}{w_d} \cdot 100\%$, where w_d and w_s are weight of dry and swollen membrane respectively.

2.3.4. FTIR spectroscopy. FTIR spectroscopy analysis of dried samples was performed with Infrared Fourier Vacuum Spectrometer Vertex 70 in 590 – 3500 cm^{-1} range.

3. Results and discussion

3.1. FTIR analysis

Experimentally acquired FTIR spectra of all samples (Figure 1.) shows several peaks that are related to Nafion® structure: e.g., at 806 cm^{-1}, what characterized S-H deformation vibrations; at 1210 cm^{-1} to C-F asymmetric vibrations; at 1152 cm^{-1} to C-F symmetric vibrations; at 1056 cm^{-1} to $-SO_3$ vibrations; at 983 cm^{-1} to C-O-C vibrations. Non-activated Nafion®/PANI sample shows 4 additional peaks compared to Nafion® sample: at 1310 cm^{-1}, 1508 cm^{-1}, 1590 cm^{-1}, 1722 cm^{-1}, which corresponds to C-N and C=C vibrations, benzenoid and quinoid forms respectively. It proves that in this sample the PANI is in emeraldine form [7]. The peak at 1643 cm^{-1} characterize C=O amide vibrations, 3500 cm^{-1} - NH^{2+} vibrations in PANI. Membranes with more intensive 1508 cm^{-1} peak designate that composite polymer has more reduced fragments (influence of hydrogenation) and vice versa. An increase of intensity of peaks at 1643 and 3500 cm^{-1} proves the fact that

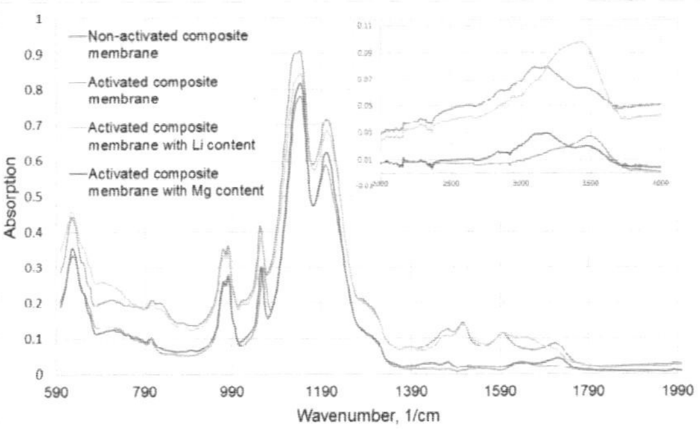

Figure 1. Composite membrane Nafion/PANI FTIR spectrum: non-activated and activated.

polyaniline had changed its structure: these peaks are characteristic to pernigraniline form of PANI, so some of the fragments of emeraldine salt form have hanged to non-conductive fragments [5,7]. Non-activated membrane has double peak at 806 cm^{-1} and 827 cm^{-1} and peak at 1390 cm^{-1}; vibrations at 827 cm^{-1} and 1390 cm^{-1} belongs to NO_3- anion.

Composite sample with magnesium content has additional peak at 1482 cm^{-1} what proves the presence of PANI in emeraldine form. Non-activated membrane with lithium content has peak at 1510 cm^{-1} what characterized NH_3^+ or N-H deformation vibrations in PANI.

Activated membrane doesn't show PANI additive because after activation process, there was a hydrogenation, what has influence on PANI state; also, hydrochloric acid can dissolve polyaniline. Some additional researches is needed to determine significance of influence of all the mentioned factors.

Comparing non-activated membranes FTIR results in the range of 2700-3700 cm^{-1} with Nafion®/PANI composite samples that have been treated with reduction ions, sample which include Li^+ ions part shows emeraldine form, but sample with magnesium ions has the similar line as non-activated membrane.

3.2. Water uptake

According to observed water absorption dynamics (Figure 2.) it is evident that 2 days is enough to reach water absorption maximum for all membranes, since obtained results after 2 and 6 days (2880 and 8660 minutes respectively) are the same.

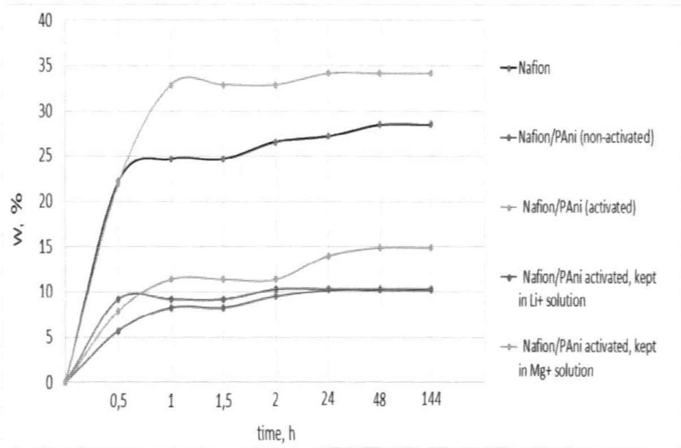

Figure 2. Water absorption kinetics results.

Compared to pure Nafion® the highest absorption ability was shown by activated Nafion®/PANI composite polymer - it absorbed 6 w% more water compared to pure Nafion® and 24 w% more than inactive composite sample. Compared to non-activated membrane samples which were treated in lithium and magnesium ion solutions, absorbed about the same and slightly more water (by 0.5 w% and 4.9 w% respectively). This could lead to increase mechanical stability due to less swollen membrane and geometrical size increase, but could cause proton conductivity decrease.

3.3. Thermal analysis

According to TGA/DSC results for pure Nafion® membrane during heating three obvious processes can be distinguished: evaporation of absorbed water which lasts to 110°C; elimination of linked water what lasts to 205 °C; degradation interval ~210-360 °C designate destruction of sulfonic group; Nafion®'s full degradation is observed before 500 °C; glass transition temperature on DSC curve is observed at ~117°C.

Even though PANI degradation happens at higher temperatures compared to Nafion®, all obtained composite membranes with polyaniline content degrade at lower temperatures. Only for lithium sample degradation starting interval was shifted to higher temperature side, but it has faster speed of degradation with nearly vertical drop of mass to nearly fully degraded sample (Figure 3). Activated and non-activated samples shows endothermal peak at ~140-145 °C on DSC curve, what mean that has started glass transition process into polymer. Although, membranes that were treated in reducer's solution show improved results compared to pure Nafion® membrane: for the composite membrane glass transition temperature is at ~140 °C.

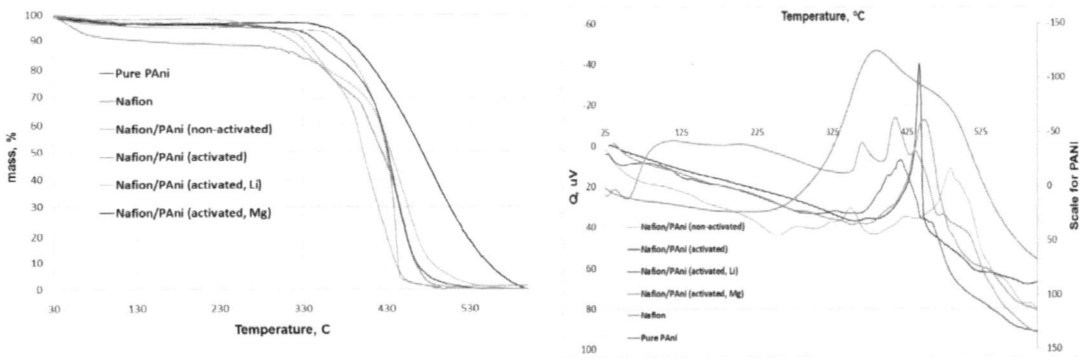

Figure 3. Composite sample TGA/DSC results.

Composites containing PANI significantly differs from pure Nafion®'s destruction. Process is characterized by noticeable decrease in losses of weight at stages of degradation of sulfonate groups and their destructions. Also end of stages is displaced in the range from 25 to ~ 40 °C. Results of membranes that were treated in reducing solution are more positive than those in which reducing agent is not active.

From obtained results it appears that in fuel cell where risk is possible when temperatures can rise above operating temperature of fuel cells, Nafion®/PANI composite membrane that is modified with lithium solution is best suited for use. It has ability to hold water in wide range of temperatures (up to ~ 120 °C) and "survive" for some time at elevated temperatures, what provide PANI content into composite polymer.

3.4. Proton conductivity

Results of original membranes were compared to results of pure Nafion® membrane. Conductivity of each sample at 100% humidity varies considerably. Adding PANI to the Nafion® causes material to loose in conductivity, but activation process allows achieving conductivity values that are comparable to the ones of the Nafion®.

Treating composite membrane with Li ions containing solution lead to the best results observed between measured samples. On the opposite, treating with Mg ions containing solution decreased proton conductivity; this sample showed the lowest conductivity value from the all samples. Probably this is caused by one or several ion's properties, because Li ion properties (size, nucleus charge etc.) are closer to the one of the hydrogen that participates in proton conductivity. Smaller size could lead to deeper Li ion impregnation; causing more homogeneous distribution in membranes volume. Reducing ions reduce quinone diimide to p-phenylenediamine on PANI and reduce number of electrostatistic bonds between Nafion and PANI, what increases protons conductivity.

Table 1. Conductivity of pure Nafion® and composite membranes - non- and activated.

Samples	σ, mS/cm
Pure Nafion®	118
Nafion®/PANI non-activated	69
Nafion®/PANI activated	145
Nafion®/PANI activated with Li^+	155
Nafion®/PANI activated with Mg^+	29

During several measurements that lasted two hours, no membrane showed any reduction in conductivity and no visible damage was observed after measurements that would indicate low mechanical or chemical durability in short time interval.

4. Conclusions

The modification of Nafion® membranes by PANI and reduction ions shows better properties than pure Nafion® sample and composites Nafion®/PANI without treatment by reduction ions. These modified membranes were characterized by impedance spectroscopy and TGA to confirm the ions positive influence on composite. Impedance spectra data demonstrated that composite modify with Li^+ ions increase proton conductivity, FTIR shows that reducing ions have influence on PANI oxidation degree. It is believed that the polyaniline particle mostly occurs in membrane`s volume and has influence on water absorption because formed polyaniline particle size in Nafion® matrix isn't limited to pore size. We conclude that composite Nafion®/PANI modification with positively charged ions is an effective way to reduce the rate of water transport through the membrane. Because of the physical electrostatic interaction between the positively charged polymer and negatively charged sulfonated groups in the membrane, the pore size shrunk.

In order to improve conductivity results for samples that were treated in reducing solution, experiments should be performed to determine optimum concentration of reducing agent (Li or Mg ion concentration in given solution) for known geometric parameters of the membrane and to find optimal elongation times to prevent such problem that ions can bind to sulfonate groups.

The results demonstrate that the content of polyaniline in a composite polymer matrix is suppressing degradation and destruction processes of sulfonate groups. Lithium ions in composite are improving membrane conductivity. This both component combination allows to get composite Nafion®/PANI with better options than the ones of the pure Nafion® polymer.

Acknowledgement
Funding from European Union`s Horizon 2020 Research and Innovation Program project under grant agreement No 768789 is greatly acknowledged.

References

[1] Devanathan Ram 2008 *Energy Environ. Sci.* **1** 101–19
[2] Napoli L, Lavorante M J, Franco J, Sanguinetti A and Fasoli H 2013 *J. New Materials Electrochem. Systems* **16** 151-6
[3] Ansari Reza and Keivani M B 2006 *E-Journal of Chemistry* **3** (4) 202-17
[4] Berezina N P, Kubaisy A A-R, Timofeev S V and Karpenko L V 2006 *J. Solid State Electrochem.* **11** 378
[5] Wasu M B and Raut A R 2015 *Int. J. Chem. Sci.* **13** 1285-94
[6] Quillard S, Louarn G and Lefrant S 1994 *Phys. Rev. B* **50** 12496
[7] Jamadade V S, Dhawale D S and Lokhande C D 2010 *Synthetic Metals* **160** 955–60